The Open University

THE HANDLING OF EXPERIMENTAL DATA

Prepared by the Science Foundation Course Team

THE OPEN UNIVERSITY PRESS

The Open University Press,
Walton Hall, Bletchley, Buckinghamshire

First Published 1970.   Reprinted 1971

Designed by the Media Development Group of the Open University.

Printed in Great Britain by
EYRE AND SPOTTISWOODE LTD
AT GROSVENOR PRESS PORTSMOUTH

SBN 335 02029 1

Open University courses provide a method of study for independent learners through an integrated teaching system, including textual material, radio and television programmes and short residential courses. This text is one of a series that make up the correspondence element of the Foundation Course in Science.

The Open University's courses represent a new system of university level education. Much of the teaching material is still in a developmental stage. Courses and course materials are, therefore, kept continually under revision. It is intended to issue regular up-dating notes as and when the need arises, and new editions will be brought out when necessary.

Further information on Open University courses may be obtained from The Admissions Office, The Open University, P.O. Box 48, Bletchley, Buckinghamshire.

1.2

Contents   Page

Introduction   1

A note about authorship of this text

This text is one of a series that, together, constitutes *a component part* of the Science Foundation Course. The other components are a series of television and radio programmes, home experiments and a summer school.

The course has been produced by a team, which accepts responsibility for the course as a whole and for each of its components.

The Science Foundation Course Team

M. J. Pentz	(Chairman and General Editor)
F. Aprahamian	(Editor)
A. Clow	(BBC)
P. A. Crompton	(BBC)
G. F. Elliott	(Physics)
G. C. Fletcher	(Physics)
I. G. Gass	(Earth Sciences)
L. J. Haynes	(Chemistry)
R. R. Hill	(Chemistry)
R. M. Holmes	(Biology)
S. W. Hurry	(Biology)
D. A. Johnson	(Chemistry)
A. B. Jolly	(BBC)
A. R. Kaye	(Educational Technology)
J. McCloy	(BBC)
J. E. Pearson	(Editor)
S. P. R. Rose	(Biology)
R. A. Ross	(Chemistry)
P. J. Smith	(Earth Sciences)
F. R. Stannard	(Physics)
J. Stevenson	(BBC)
N. A. Taylor	(BBC)
M. E. Varley	(Biology)
A. J. Walton	(Physics)
B. G. Whatley	(BBC)
R. C. L. Wilson	(Earth Sciences)

The following people acted as consultants for certain components of the course:

J. D. Beckman	B. S. Cox
H. G. Davies	R. J. Knight
D. J. Miller	M. W. Neil
J. R. Ravetz	H. Rose

EXPLANATORY NOTE

How to use "The Handling of Experimental Data" (HED)

HED is meant to be for reference when needed. It is *not* meant to be read "in one go", as the main text of a Unit might be.

You are *not* expected to learn everything in *HED*. However, *some* of the terms, concepts and principles in *HED* are listed in column 3 of Table A at the beginning of the main text of some Units — e.g. Units 1 and 2. This means that they will be included in Objective 1 of the list of objectives for these Units.

Some of the information in *HED* will not be needed at all in this course — for instance, many of the units in section 3. We included it for the sake of completeness and because you may find it useful in later courses.

You will usually be advised in the main text when you should consult *HED*. (See, for example, pages 29, 32 and 36 of Unit 1.) Otherwise you should do so whenever it seems to you that it might be helpful. You will naturally be able to judge this better if you have a general idea of the contents. So we advise you to skim rapidly through the handbook at an early stage in the course.

Introduction

One cannot expect to get a "feel" for science by just reading about it, so most science courses involve the student in *doing* things — our Foundation Course is no exception.

As you will appreciate it is not easy to involve students of the Open University in experimental work. The fact that we are making sure that our students are equipped with what amounts to their own laboratories should leave you in no doubt as to the extreme importance we attach to this aspect of your work. The quality of your experimental work will go a long way to deciding your continuous assessment rating. But why do practical work?

In Unit 1 and Unit 2 you can read how important are observation and measurement to the scientist. These provide the raw material from which he is able to build up his understanding of the world. But the gathering and interpretation of data is fraught with difficulty. There are many ways in which data can be faulty and misleading. Great skill, care, common sense, and ingenuity are necessary to produce good measurements. We want you to experience at first hand some of the enjoyment and sense of achievement that can go with the making of such measurements, and to share with the scientist his disappointments and frustrations when experiments go wrong and observations are hard to interpret.

Through practical work you will learn how to handle data — how to convert raw observations into an estimate of the value of some physical quantity. You will learn how to assess the precision of your result. Phenomena you have read about will be more clearly understood and more easily remembered when you have actually seen them for yourself. We hope you will develop critical awareness, a certain manual dexterity, a familiarity with some pieces of scientific equipment, and so on. And if you are one of those going on to a science-based career, you will acquire a necessary measure of self-confidence. One day you may find yourself having to persuade your head of department to effect a policy change as a result of some experiment you have performed. To do this you will need a certain measure of confidence in your own ability. Likewise if you *are* the head of the department you will need confidence in your own judgement when assessing the reliability of the work of a possibly over-enthusiastic young employee.

This handbook (which is referred to throughout the course as *HED*) is divided into four sections. The first is a description, in very general terms, of some of the ways in which scientific data can be faulty. In it you will see, for example, how the problems facing a physicist can differ quite markedly from those facing a biologist or astronomer. The purpose of this section is to place your practical work in context. We want you to be aware, while working at home, that scientists investigating current problems are facing exactly the same types of difficulty as you are. The subjects being studied may differ, but the types of difficulty and the skills required to overcome them are the same.

The second section is quite different in character. It is a down-to-earth guide to practical work — in fact, a manual. It contains a good many tips we hope you will find useful. These are summarized in a 50-point Check List which you should keep referring to throughout the year. The advice offered in it applies to your experimental work in general and is intended to supplement the detailed instructions issued with individual experiments.

The third and fourth sections deal with units and dimensions respectively.

Section 1

Limitations on the Accuracy of Experimental Data

The scientist must work with reliable data. In this section, we describe some of the various ways faulty results can arise.

1.1 Human error

This may be due to rank carelessness, where perhaps a scale is simply misread. Repeated readings of the same quantity often reveal this type of mistake.

More difficult to detect is an error in technique which leads to all the repeated readings being biased in the same way. A common fault of this type is *parallax*. This occurs when reading, for example, the position of a pointer on a scale (e.g., on a stop-watch). Figure 1a shows such a pointer as seen from above, Figure 1b as seen end-on. Clearly the scale reading R_c is the correct one and in order to make it the eye must be placed directly above the pointer at E_c. At any other position, say E_w, a wrong reading, R_w, is taken.

With care and experience one becomes more adept at avoiding errors such as this. The design of instruments can also help in this respect. In order to avoid parallax, for example, many dials incorporate a mirror alongside the scale — by positioning the eye so that the pointer and its reflection are superimposed, you know that the eye is directly above the pointer and correctly placed for taking the reading. To reduce still further the chance of making a reading error, instruments these days often display their reading directly as a number and not as a position of a pointer (you will have noticed that clocks of this type are increasingly being preferred for public buildings).

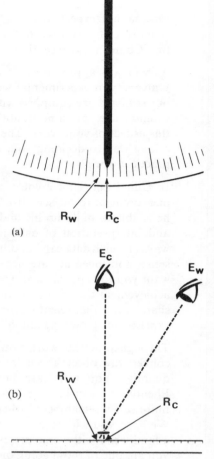

(a)

(b)

Figure 1   Avoiding parallax error

1.2 Instrumental limitations

Instruments have their own inherent limitations. Some are obvious, as in the case of cheap wooden rulers where one can see by eye that the divisions are not evenly spaced. More sophisticated pieces of equipment however may be subject to several sources of error. Take for example a microscope. Although microscopes are most often used simply to view small objects, it is sometimes necessary to measure the size of the object. The accuracy of such measurements depends upon a number of factors — the rigidity of the column supporting the lenses, the rigidity with which the specimen is clamped to the stage, and the accuracy with which the divisions have been engraved on the scale. Some microscope stages are moved by turning a calibrated screw, each complete turn corresponding to a certain movement of the stage (Fig. 2). The accuracy of the measurements made on such instruments depends upon the uniformity of the screw thread. It also often happens that after the stage has been moved some distance in one direction it does not immediately respond when the motion of the screw is reversed. This is called back-lash. It leads to readings that differ, depending upon the direction from which the microscope approached the position.

How could you overcome this particular trouble?

Figure 2   A calibrated screw

This is simply remedied by always approaching a position to be measured from one chosen side only.

Of course, instruments vary markedly in quality; some are better designed and incorporate better workmanship than others. This does not mean to say one should always go for the most accurate and precise equipment — for one thing it is liable to be expensive!

Indeed a scientist often has to assess in advance how good his instruments need to be in order to give him sufficient accuracy for the task in hand (see for example the discussion of Galileo's experiment in Unit 1). This is something that is not peculiar to a scientist — the man behind the counter at the post office generally has two weighing scales, a small one for weighing airmail letters accurately, and a more robust one for weighing large parcels.

In your home-kit experiments, some of the equipment will be fairly simple. This is inevitable, as we have a large number of students to supply and need to keep down costs. We hope in fact to turn this situation to your advantage. Too often in a conventional teaching environment, a student handles only relatively sophisticated apparatus — apparatus whose reliability he may hesitate to question. You, on the other hand, by the end of the year will have had instilled in you a very healthy scepticism for the performance of any equipment! This we hope you will always retain. Naturally, at some point the student specializing in science must be given the chance to handle more refined tools — this chance we hope to give you at summer schools and in later years.

1.3 Extraneous influences — the need to exercise control

Often an experiment goes wrong because the effect being measured is caused by something other than that assumed. For example, the reading on a balance depends not only on the weights in the pan, but also on whether there is a draught in the room. The measurement of the length of an object depends not only on estimating which mark on the steel ruler is the appropriate one, but also on whether the temperature at the time of making the reading was the same as that when the ruler was inscribed — if not, the overall length of the ruler has changed. A chemist investigating the properties of a substance may find his conclusions completely invalidated by the presence of a tiny trace of impurity in the sample under test. Hence, in all quantitative chemical work the purification of the materials is a routine requirement, unless a satisfactory level of purity is guaranteed by a reputable manufacturer.

When performing practical work, one must always be on the lookout for extraneous influences that might affect the results. Some must be eliminated — like shielding the balance from draughts, or meticulously cleaning beakers so as not to contaminate samples with impurities. For others, it may be easier to calculate the effect they produce. For example, if the temperature of the ruler had been taken at the time it was inscribed, and if the temperature at the time of measurement could be kept constant, it would be possible to calculate the change in length from the known properties of steel.

You will begin to see that in order to carry out really good scientific work a certain control has to be exercised over the environment. Thus a scientist tends to work whenever possible in a special environment where the conditions he desires may be realized — this environment is called a *laboratory* (see the TV programme of Unit 1). Just as the layout of a kitchen or a bathroom in a house is determined by the functions they serve, so a laboratory is purpose-built and its layout is geared to the type of condition to be established in it.

The idea of excluding unwanted effects extends not only to influences that have nothing to do with the subject under investigation (like draughts and impurities), but also to the influences that are inescapably part of the phenomenon being studied.

For example, suppose you were studying the transmission of heat through solids. It is common experience that some substances transmit heat better than others — hence the handle of a kettle is non-metallic. But suppose you wanted to investigate in detail the heat-conducting properties of the material from which the handle of the kettle is made — how would you go about it?

Well, for a start, you would not want the material in such a complicated form as a handle of a kettle. You would want simpler shapes — circular discs of uniform thickness perhaps. Then you could try the effect of keeping the temperature difference across the faces constant and, for faces of the same area, investigate how the heat flow depended on the thickness of the disc (Fig. 3a). Next, keeping the thickness constant, you could try the effect of varying the area (Fig. 3b). Finally, keeping both area and thickness constant, you could see what happened when the temperature difference was altered.

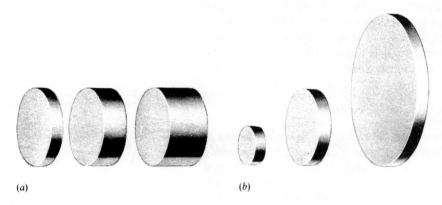

(a)   (b)

Figure 3   (a) Discs of constant area and variable thickness; (b) Discs of constant thickness and variable area

Thus, the dependence of the heat flow on either the area, or the thickness, or the temperature difference, could be investigated by keeping the other two fixed. This then is another kind of "control" a scientist may exercise; it is a control over factors that are integrally part of the subject under investigation. He cannot "eliminate", for example, the area of the disc, in the same way as he could eliminate a draught or impurity, but he can keep the area constant while he investigates how the rate of heat flow alters in response to changes in some other quantity, for example, the thickness of the disc.

Such procedures do, however, assume that the subject being studied lends itself to this kind of control — in the case we have been describing, that the same material is available in the more conveniently shaped discs. This is usually so in the fields of physics and chemistry, where it is easy enough to get hold of convenient quantities of materials and chemicals.

The biologist is not so fortunate. He is generally concerned with systems of enormous complexity. Even with relatively simple organisms, it is difficult to isolate and study separately the various influences affecting behaviour. Indeed, some branches of the subject are specifically concerned with the behaviour of animals in their natural environment, i.e., subject to no artificial constraints. However, biology today is becoming increasingly concerned with the behaviour of single-celled organisms like bacteria or viruses, or *portions* of organisms — cells or fractions of cells, isolated muscle fibres or nerves etc. Operating on these, the fields merge to give rise to biochemistry and biophysics (see, for instance, Units 14 to 17).

The earth scientist contends with such large-scale phenomena that his ability to control extraneous influences is exceedingly limited. For example, when studying erosion by wind and rain, he is of course quite unable to switch off, say, the wind while he studies the effect produced by the rain alone. When studying the age of a mountain range, he is in no sense able to control the situation — he must work from his observations of the situation as he finds it.

This is not to say that the earth scientist must always play the role of a passive observer. In certain instances, he is able to re-create a given feature of nature in a laboratory. For example, you will have noticed that the surface of the sand on a seashore often has ripples in it. Such patterns can be reproduced in an apparatus called a flume tank. In it, as you will see in Unit 26, one can vary at will the rate of flow of water, the depth of the water, and the size of the sand grains, and in this way study the patterns produced under a variety of conditions. Also, the earth scientist depends to a great extent on sophisticated techniques developed in laboratories, e.g., those involved in the radiometric dating of rocks and in the recording of earthquakes.

The astronomer has no control over the stars he studies — he is a purely passive observer. Until recently he had even to make his observations through a turbulent and partly opaque medium — the Earth's atmosphere. This difficulty could be partly offset by taking his instruments to the tops of mountains or by flying them in balloons; now of course he can expect to make increasing use of satellites orbiting above the atmosphere.

1.4 Unrepresentative samples

No matter how carefully a scientist makes his observations, if the sample being tested is unrepresentative he is wasting his effort.

For example, a geologist may decide to pick up only complete fossils. This may seem a perfectly reasonable thing to do, because such fossils can be examined more thoroughly. However, the unbroken fossils may not be representative of fossil assemblage as a whole; they may have some special characteristic that has reduced their chances of being broken. Likewise, when required to determine the abundance of various types of rock present in a given area, he may be attracted by the more unusual types. He may tend to collect the exotic and spectacular specimens, in preference to ones that are more abundant, drab and uninteresting.

Or we can take the case of a physicist studying the ways in which nuclear particles of one type deflect on hitting particles of another type. The arrangement of his equipment may be such that he has less chance of detecting a collision when the incident particle is scattered through a small angle than when it is scattered through a large one. The sample of collisions he records is thus not truly representative of all the collisions taking place.

1.5 Statistical fluctuations

Even when care has been taken to exclude biases, the sample being studied is still subject to statistical fluctuations.

If, for example, there were a 2 per cent chance of a nuclear particle scattering through an angle lying between $10°$ and $15°$, this does not mean to say that when the physicist has detected 10 000 collisions he will find exactly 200 lying within this angular range; he may find only 180, or perhaps 212. In these cases his best estimates of the chance would be 1.8 per cent and 2.12 per cent. (In Figure 36 of Unit 2 you can find a further example of the effects of statistical fluctuation. For instance, after a time of $1\frac{1}{2}$ half-lives the "expected" number of nuclei left, as read off from the graph, was 57; the actual number 64.) The only way to reduce statistical errors and increase the precision of the final result of such an experiment is to increase the size of the sample.

1.6 Disturbances caused by the act of observation

Finally, let us mention that the very act of the scientist making an observation can alter the situation being studied.

This is particularly true of biological investigations. As soon as an attempt is made to study the behaviour of an animal under controlled conditions, the mere fact that it is being observed may affect its behaviour.

If the pressure in a tyre is being measured, the attachment of the pressure gauge releases some of the air from the tyre and so alters the pressure being measured.

The immersion of a cold thermometer in a beaker of hot liquid lowers the temperature of the liquid being read. If the thickness of a sheet of paper is measured with a micrometer screw-gauge, the pressure exerted by the jaws of the gauge squashes the paper and so reduces its thickness.

Indeed, strictly speaking *no* observation can be made without, to some extent, altering what is being observed. Even when, for example, an object is simply illuminated in order to look at it, the light exerts a force on it — admittedly an exceedingly small force — and this force very slightly alters the state of the object being viewed. It should be emphasized that this effect is so minute as to be completely negligible in the context of normal everyday experience. But in the study of the behaviour of the smallest constituents of matter, it is found that even these small forces assume enormous importance — so much so that the very nature of observation has to be called into question. We take up this problem in Unit 29.

A Guide to Practical Work

From what you have just read you will appreciate that *a significant aspect of the work of an experimental scientist is concerned with identifying sources of error, reducing their effects, and assessing the reliability of his final result*. Much of what we have to tell you in this second section is to do with the handling of errors; indeed, the first six sub-sections are devoted to this topic. We shall have to quote you a number of formulae based on the theory of statistics. This in a way is unfortunate; we do not like producing a formula out of thin air and telling you simply to take our word for it. But in the present case it is unavoidable; you would need to take a mathematics course to appreciate how some of the formulae that follow are derived. So we ask the non-mathematically inclined student to be content to use them as *tools*. At least we can assure you that you are not expected to remember them all; there is nothing to stop you referring to this handbook throughout the year — indeed you are encouraged to do so.

2.1 Repeated readings; estimation of the mean value

Never be satisfied with a single reading, no matter what measurement you are making — *repeat it*. This procedure will improve the precision of your final result; it can also show up careless mistakes. All measurements are subject to *random errors* and these spread the readings about the true value. Sometimes the reading is too high, sometimes too low. With repeated readings, random errors tend to cancel out.

As an example, suppose you are measuring the volume of water flowing through a tube in a given time. Five readings of this quantity may yield the values:

$$436.5, 437.3, 435.9, 436.4, 436.9 \text{ cm}^3$$

Assuming that there is nothing to choose between these values (for example that you did not concentrate much harder for the last reading than you did for the first four), then the best estimate for the volume of water is the average, or mean value, of the five readings, i.e.

$$\bar{x} = \frac{436.5 + 437.3 + 435.9 + 436.4 + 436.9}{5} = 436.6 \text{ cm}^3$$

(If the last measurement *was* more reliably made than the others then of course you should go back and do all of them that carefully!)

In general, if n readings, $x_1, x_2, \ldots x_n$ are taken of a single physical quantity, and the readings are taken under the same experimental conditions (i.e., with the same care and reliability), then the best estimate is the *mean*, $\bar{x}$, given by:

$$\bar{x} = \frac{x_1 + x_2 + x_3 \cdots + x_n}{n} \quad \ldots \ldots \ldots \ldots (1)$$

(In the example we considered, n was five.)

2.2 Distribution of measurements; random errors on the mean value

It is no good just quoting a result and leaving it at that. Someone else may wish to use your result and will want to know what reliance to place on it. Suppose, for example, he wishes to know the volume of water issuing from the tube. He has your value (the value 436.6 cm³ derived above) and perhaps another value of 436.0 cm^3 from a second experimenter. He wants the best possible value. He cannot choose one and ignore the other, for he has no grounds for making such a choice. Does he take the mean? I.e.,

$$\frac{436.6 + 436.0}{2} = 436.3 \text{ cm}^3$$

If the second experimenter had taken only one reading, *viz*, 436.0 cm^3 you would naturally feel upset that his single reading should carry as much weight as the mean of your *five* readings.

If the reading taken by the second experimenter had the same reliability as any of the readings you took, what *would* be the best estimate in these circumstances?

$$\bar{x} = \frac{436.5 + 437.3 + 435.9 + 436.4 + 436.9 + 436.0}{6}$$

$$= 436.5 \text{ cm}^3$$

Alternatively, of course, his value of 436.0 cm^3 may have been the average of more readings than you took.

If it were known that the second experimenter had taken 20 readings of the same reliability as yours what would be the best estimate?

There is no need to know what the 20 individual values were. 436.0 is the average of 20 readings and 436.6 the average of 5. Therefore,

$$\bar{x} = \frac{(436.0 \times 20) + (436.6 \times 5)}{25}$$

$$= 436.12 \text{ cm}^3 \approx 436.1 \text{ cm}^3$$

Clearly the number of readings taken plays a part in determining the precision of a result. Is it the only factor?

Take a look at Figure 4. In Figure 4a are displayed the five readings that gave a mean value of 436.6 cm^3. In Figure 4b are five other possible readings giving the value 436.0 cm^3. If the number of readings were the sole criterion for judging the precision of a result, the two values, 436.6 and 436.0 cm^3, would carry the same weight. Does it look from Figure 4 as though they ought to? It would seem not. The measurements belonging to the second set appear to agree with each other more closely. It may have been that the second experimenter had a better way than you of cutting off the water flow at the end of the given time period. It is only right that he should be given credit for his extra care. *The extent to which the readings are spread out about the mean position has to be taken into account.* A quantitative measure of this spread can be found in the following way:

(a)

(b)

Figure 4   A set of five measurements of a volume; (a) with a wide spread; (b) with a narrow spread

First calculate the mean, $\bar{x}$, from equation (1). Then calculate the *residuals*, $d_1 \ldots d_n$, which are defined simply as the difference between the individual readings and the mean

$$d_1 = x_1 - \bar{x}$$
$$d_2 = x_2 - \bar{x}$$
$$\vdots \qquad \cdots \cdots \cdots \cdots \cdots (2)$$
$$d_n = x_n - \bar{x}$$

The *standard deviation of the sample*, s, is then defined by the equation:

$$s = \left\{ \frac{(d_1^2 + d_2^2 + \cdots + d_n^2)}{n} \right\}^{\frac{1}{2}} \quad \cdots \cdots \cdots \cdots (3)$$

and s is the required measure of the spread of the readings about the mean value.

Calculate s for the original five readings for the volume of water (i.e., 436.5, 437.3, 435.9, 436.4 and 436.9 cm³).

If you wish to have further practice in calculating a standard deviation, try self-assessment question 1, p. 44.

The working is as follows:

$$d_1 = 436.5 - 436.6 = -0.1$$
$$d_2 = 437.3 - 436.6 = 0.7$$
$$d_3 = 435.9 - 436.6 = -0.7$$
$$d_4 = 436.4 - 436.6 = -0.2$$
$$d_5 = 436.9 - 436.6 = 0.3$$
$$\therefore \quad d_1^2 = 0.01$$
$$d_2^2 = 0.49$$
$$d_3^2 = 0.49$$
$$d_4^2 = 0.04$$
$$d_5^2 = 0.09$$

Total $\qquad = 1.12$

$$s^2 = \frac{1.12}{5} = 0.224$$

$$\therefore \ s = 0.47 \text{ cm}^3$$

or $s \approx 0.5 \text{ cm}^3$

If s has been determined for a sample consisting of a great many readings, it gives a measure of how far individual readings are likely to be from the true value. It can be shown from statistical theory that about 68 per cent of the readings will lie within $\pm s$ of the true value, 95 per cent within $\pm 2s$, and 99.7 per cent within $\pm 3s$.*

If there are many readings, it is difficult to make a quick assessment of the precision of an experiment and the likely mean value by examining a long list of numbers displayed in a table. It is more convenient to plot the readings in the form of a *histogram* (Fig. 5).

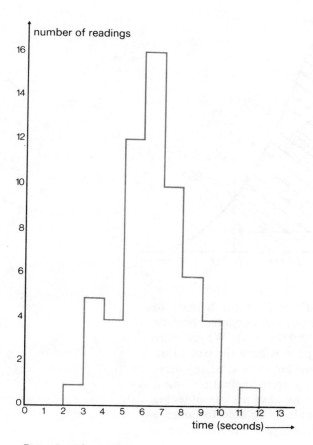

Figure 5 A histogram

The range of measured values is divided into equal intervals and a note made of the number of readings falling within each interval. Some readings may occur on the boundary between two intervals e.g. if the intervals are 0.0 to 2.0 cm, 2.0 to 4.0 cm, 4.0 to 6.0 cm, etc., you may have a reading of 2.0 cm. In which of the two adjoining intervals should you record it? One possibility is to record "half an event" in each, the other is to place it in the higher interval, remembering to put the next borderline reading into the lower of its two adjoining intervals, the next in the higher, etc.

These numbers are simply quoted to you. They assume the distribution to be a Gaussian distribution. This is black-page material and we refer the interested student to Practical Physics *by G. L. Squires, Chapter 3, or almost any other book on errors and statistics.*

The number of readings in each interval determines the height of the histogram for that interval. For instance, in the histogram shown in Figure 5 the number of readings between 6 and 7 seconds is 16. With such a histogram it is possible to see at a glance that the mean value is about 6.5 seconds and that the average total spread of the readings (corresponding to 2s) is roughly 4 seconds.

As the number of readings is increased so one can take smaller and smaller intervals and still have a reasonable number of readings in each. Eventually the "step-like" character of the histogram is no longer noticeable and one is left with a smooth curve as in Figure 6. Strictly speaking, it is only when there are very large numbers of readings that one can say: 68 per cent of the readings will be within $\pm s$ of the true value, and 95 per cent within $\pm 2s$.

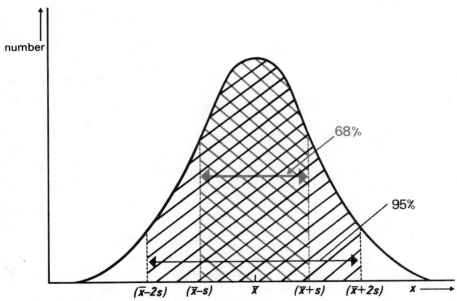

Figure 6   Distribution of a large number of readings

So much for the question of how far an *individual* reading is likely to be from the true value. This, however, is not our main concern, because normally you will take several readings and find the mean. We are more interested in how far the mean of the readings, $\bar{x}$, is from the true value. Of course it is not possible to say exactly how far away it is, any more than it is possible to say exactly how far away an individual reading is from the true value. We must emphasize that we have no way of telling what the true value is. However, it is desirable to be able to assign a probability to the mean lying within a certain range of the true value.

This range will depend not only on the spread of the individual readings but also on n, the number of such readings. It is, in fact, specified by a quantity called the *standard error on the mean*, s_m, and is estimated through the equation

$$s_m = \frac{s}{(n-1)^{\frac{1}{2}}} \dots\dots\dots\dots\dots\dots(4)$$

It is such that the mean value of a given sample, e.g., $\bar{x}$, has a 68 per cent chance of lying within $\pm s_m$ of the true value, 95 per cent within $\pm 2s_m$, etc.* s_m therefore is a measure of how close the mean value of the given sample, $\bar{x}$, is to the unknown true value. Once again we must stress that, unless you are prepared to take a course in statistics, this is an equation you must simply accept and use as a tool.

Strictly speaking the standard error is not defined through equation 4; the equation gives a way of estimating its value for a given sample. Once again refer to a book on statistics, for example, Squires, op. cit. for further details.

In the experiment we have been considering, $\bar{x}$ is the mean of the five readings, and has a value 436.6 cm³. Now calculate the standard error on it, using the value you have already calculated for s, i.e. 0.47 cm³.

For further practice, now try question 2, p. 44.

For this problem n in equation 4 has the value 5, and s is 0.47.

$$s_m = \frac{0.47}{(5-1)^{\frac{1}{2}}} = \frac{0.47}{2}$$
$$= 0.235 \approx 0.2 \text{ cm}^3$$
$$\therefore \bar{x} = (436.6 \pm 0.2) \text{ cm}^3$$

Note that the unit, in this case cubic centimetres, applies both to the mean value and its error. Therefore we have included both these numbers in the brackets.

Let us reiterate that, just because one assigns a standard error to a result (for example, ± 0.2 cm³ on a measurement of 436.6 cm³), it does not mean to say that the true value necessarily lies within the specified range (i.e., within the range 436.4 to 436.8 cm³) as would be the case if we were specifying an engineering tolerance. *It is a convention* — one that is understood by all scientists to mean that there is about a 70 per cent chance of the true value lying in that range.

Note from equation 4 that s_m goes down in value as the number of readings in the sample, n, increases — hence the desirability of repeated readings. However it does so only slowly because n appears as a square root. Thus a ninefold increase in the number of readings produces only a factor 3 improvement in precision. A large number of sloppy readings therefore is no substitute for taking care and keeping the spread of the readings, and hence s, as small as possible.

If there are many readings, it can be tiresome to have to calculate the squares of the residuals in order to obtain s from equation 3. Here is a quick alternative. It is only an approximation but it is often good enough: calculate the mean value, r, of the moduli* of the residuals, i.e.

$$r = \frac{|d_1| + |d_2| + \cdots + |d_n|}{n} \quad \ldots\ldots\ldots\ldots (5)$$

Then calculate s from the simplified expression

$$s = 5r/4 \quad \ldots\ldots\ldots\ldots\ldots (6)$$

For the experiment being considered calculate s for the five readings using equation 6 and compare with the value obtained earlier by the exact formula, equation 3.

For further practice now try question 3, p. 44.

The sum of $|d_1|, |d_2|, |d_3|, |d_4|$ and $|d_5|$ is $0.1 + 0.7 + 0.7 + 0.2 + 0.3 = 2.0$ cm³.
∴ From equation 5 the mean value of the residuals $r = 2.0/5 = 0.4$ cm³.
∴ From equation 6

$$s = \frac{5 \times 0.4}{4} = 0.5 \text{ cm}^3.$$

This compares well with the previous value of 0.47 cm³.

When equation 6 is combined with equation 4 we get an approximate relation called Peters' formula:

$$s_m = \frac{5}{4} \frac{r}{(n-1)^{\frac{1}{2}}}$$

This is very easy to evaluate and you are advised to use it.

2.3 Systematic errors

From equation 4 it is obvious that by taking a sufficiently large number of readings the random errors can be made as small as one likes. Not all errors behave in this way though. As became clear in section 1, some errors cause all measurements to be systematically shifted in one direction — these are called *systematic errors*. Figure 7a shows once again a spread of readings caused by random errors; they are approximately centred about the true value. If a systematic error is also present these randomly spread readings will be shifted so that they are no longer clustered about the true value but about some other value (Fig. 7b). Under these experimental conditions no matter how many readings are taken, the final result will not home in on the true value.

*The modulus of a negative number is the number with its negative sign changed to positive; the modulus of a positive number is the number with its sign unchanged.
Thus the modulus of -2 is: $|-2| = 2$
and the modulus of $+3$ is: $|3| = 3$

11

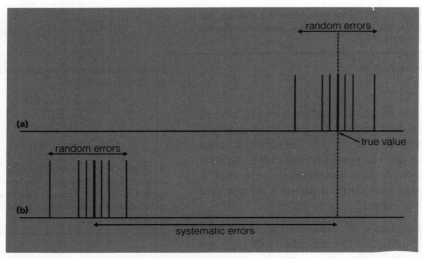

Figure 7   Random and systematic errors

It is customary to distinguish between an *accurate* and a *precise* measurement. An accurate measurement is one in which the systematic errors are small; a precise measurement is one in which the random errors are small. If both types of error are small, the measurement is spoken of as being both accurate and precise.

In contrast to random errors, which, as you have seen, may be dealt with by neat standard procedures, systematic errors are tricky, and the way they are dealt with depends largely on the skill of the experimenter. Time and again experimenters, often distinguished and very experienced ones, have produced results that disagree; always the trouble has been traced to one (or more!) of them having seriously misjudged, or overlooked, a systematic effect to which their particular experimental set-up was prone.

The first aim with systematic errors must be to keep them as small as possible. For example, as mentioned in section 1.1, when reading the position of a pointer on a scale you must avoid parallax.

Another common error is called a *zero error*. When the scale of an instrument is read, a check should always be made to see that when it should read 0, it really does point to zero. If not, the instrument must be adjusted to read zero or, if that is not possible, the zero reading has to be subtracted from all other readings. When a length is measured with a metre rule, it is best to do it between two intermediate points on the rule rather than from one extreme end — the rule may be worn at one end or the scale might not have been correctly inscribed relative to the end of the rod (Fig. 8).

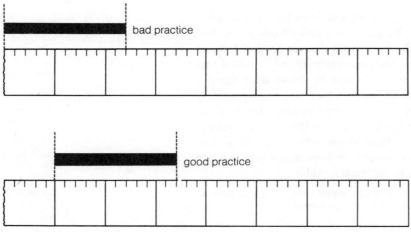

Figure 8   Avoiding systematic zero error

When the temperature of a liquid is measured, the thermometer ought to be immersed to the depth specified on it. When the instrument was originally calibrated it was immersed to this depth and, if you take a measurement under some other conditions, the expansion of the mercury in the stem of the thermometer will not be the same.

A stop-watch used for timing may run too slowly, or too fast.

Indeed the list of possible systematic errors is endless, each experimental set-up having its own particular hazards. It is not possible to give a convenient list of errors distinguishing those that are systematic from those that are random. A source of error producing a systematic bias in one type of experiment may give rise to a random error in another experiment. For example, if the masses of various objects were measured with a balance and there was in one of the pans a weight with a mass less than its face value, the measurements would be systematically biased to low values. If, on the other hand, this weight were not always used and others, some too heavy and some too light, were used as well, the errors arising would work in opposite senses to give a random effect. Thus we cannot state categorically that some sources of error always give systematic effects; each experimental set-up must be studied on its own merits.

Having taken the trouble to keep your systematic errors as small as possible, you have to estimate the size of any likely remaining error and incorporate it in the error on your final result. But how, you may well ask, can the size of the remaining systematic error possibly be estimated? After all if you *knew* of such an error you would correct for it. And of course repeated readings do not disclose residual systematic errors.

You will no doubt be relieved to know that in most cases all we can expect of you (or indeed of anybody) is that you have a rough stab at an answer.

Suppose, for example, that you have measured the length of an object using a metre rule and want to know whether the scale on the rule is accurate. You could try measuring the length using different sections of the rule. If you found that the measured length differed from one region of the scale to another by a certain percentage, then this at least would warn you that the manufacturers were not aiming at a level of accuracy better than that (or if they were aiming at it, they manifestly did not achieve it). It would be appropriate in the circumstances to take this as an estimate of the systematic error. Of course, it is not a very satisfactory estimate and ideally you would like to compare it with a scale that is known to be accurate; but it is unlikely that you have a more accurate one at home.

> **Suppose you found the object measured exactly the same length, say 31.65 cm, anywhere over the scale. When the same object was measured on another scale, this second scale was also found to be exactly uniform — but the length of the object this time was 31.70 cm. How is it possible to decide which length is right — how can one speak of one scale as being "known to be accurate"?**

Refer to Section 2.4.2 of Unit 2

Fortunately you *do* have access to a standard of time measurement in the home (or if not in the home, in the local telephone kiosk). You might like to invest in a telephone call to "the speaking clock" to check the accuracy of your stop-watch.

With regard to a zero error, although you may have corrected for it as far as possible, your ability to read a scale accurately is limited. You are not likely, for example, to be able to estimate a reading to better than one-fifth or one-tenth of a division. All readings therefore will be subject to a residual zero error of this magnitude.

These examples may give you some idea of how to tackle this difficult task of estimating (or perhaps we might more appropriately say "guesstimating") systematic errors. Certain errors are totally incapable of being expressed quantitatively. If this is the case you should simply mention the source of possible error in your account so that at least we know you have thought about it.

2.4 Combining random and systematic errors

You will recall that earlier we considered you as having done an experiment in which you measured the volume of water flowing through a tube in a given time, and the mean of five readings was 436.6 cm³. The random error on this quantity you calculated to be 0.2 cm³ from the formula for the standard error on the mean (equation 4, p. 10).

Now suppose you have made an effort to estimate the systematic errors. Perhaps you think you might have had a tendency to start the stop-watch a fifth of a second too fast, or too slow, each time. This is a common type of systematic error and is called a personal error. It arises because each individual responds in certain characteristic ways to aural and visual stimuli. (You do not *know* that you have made any such error, but it is the sort of error you *could* have made and one-fifth of a second seems a reasonable magnitude for it — anything longer you would probably have detected.) If the total length of time involved in the measurement was, for example, 4 minutes, i.e. 240 seconds, an error of one-fifth of a second would be an error of

$$\frac{1/5}{240} \quad \text{i.e. 1 part in 1200}$$

The corresponding error in the volume of water would be

$$\frac{436.6}{1200} = 0.4 \text{ cm}^3$$

When reading the level of the water collected in the measuring flask you would probably have noticed that this was not easy. The surface of the liquid was curved and there was a possible parallax error due to the liquid face not being in direct contact with the inscribed markings on the outside of the flask (Fig. 9). To cover these effects you have assigned a further systematic error of 0.2 cm³.

Thus there are three separate errors: 0.2, 0.4, and 0.2 cm³. The problem is — *how do you combine them to give a final quoted value?*

Would it be reasonable to add them — i.e., quote the final error as

0.2 + 0.4 + 0.2 = 0.8 cm³?

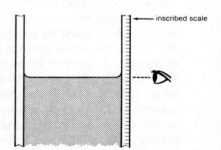

Figure 9   *Possible parallax error in reading a water level*

No. This is being unduly pessimistic! In order to reach a value of 0.8 cm³ it would be necessary for all three errors to push the measured value in the same direction, i.e., all towards an over-estimate or, alternatively, all towards an under-estimate of the volume of water. In some experimental situations this *could* happen. If, for example, the volume of a box is measured using a ruler that gives systematically high readings, then the height, breadth, and length of the box will be affected in the same way. All three contributing errors to the volume will then be directed towards an over-estimate. But if the errors are independent of each other, i.e., there is no reason why the errors should all operate in the same sense, it is more reasonable to expect some to push the estimate up and others to push it down — there will generally be some partial cancellation and this should be allowed for. This applies when compounding systematic errors as well as random errors. One error may systematically push the estimate in one direction but another may well push the estimate systematically in the opposite direction.

Yet again we must appeal to statistical theory and quote you a result:

the resultant standard error is obtained by summing the squares of the individual errors and taking the square root. Thus for independent random or systematic errors, $\pm e_1, \pm e_2, \ldots \pm e_n$, the resultant error, E, is given by:

$$E = (e_1^2 + e_2^2 + e_3^2 \cdots + e_n^2)^{\frac{1}{2}} \ldots \ldots \ldots \ldots (8)$$

Considering the errors to be independent of each other, now calculate a resultant error for the experiment you have been considering, i.e., where the contributing errors were 0.2, 0.4 and 0.2 cm³.

$$E = (0.2^2 + 0.4^2 + 0.2^2)^{\frac{1}{2}}$$
$$= (0.04 + 0.16 + 0.04)^{\frac{1}{2}}$$
$$= (0.24)^{\frac{1}{2}}$$
$$= \underline{0.5 \text{ cm}^3}$$

We said that when there are several errors, allowance must be made for partial cancellation. Does this mean that rather than have a single source of error it is better to introduce a second one of comparable magnitude in order that the two will tend to cancel each other and give a better estimate?

If both errors have a value, e, then the resultant error deduced from equation 8 is:

$$(e^2 + e^2)^{\frac{1}{2}} = (2e^2)^{\frac{1}{2}} \approx 1.4e$$

Thus the situation is worse than if only one source of error were present (which of course is only to be expected!).

2.5 Errors on sums, differences, etc.

So far we have considered errors on a single quantity. Often however the aim of an experiment is to evaluate something that has incorporated in it several measured quantities each with its own error. The problem then is — what is the error on the final calculated quantity?

Suppose you are to evaluate a quantity, Z, which is calculated from the expression

$$Z = A + B \ldots \ldots \ldots \ldots \ldots \ldots (9)$$

where A and B are measured quantities each with an error ΔA and ΔB (pronounced "delta A" and "delta B"). The error, ΔZ, on Z will arise partly from ΔA and partly from ΔB. As in the previous section, the errors are not simply added because sometimes the error on A may work in the opposite direction to that on B. Once again the square root of the sum of the squares is taken for the resultant error:

$$\Delta Z = \{(\Delta A)^2 + (\Delta B)^2\}^{\frac{1}{2}} \ldots \ldots \ldots \ldots \ldots (10)$$

If Z is represented by the difference of A and B, i.e.

$$Z = A - B \ldots \ldots \ldots \ldots \ldots \ldots (11)$$

then the same considerations apply. Although B now has a negative sign in front of it, ΔB is still just as likely to be positive as negative. So as far as the *errors* are concerned nothing has changed — their magnitudes are the same and they still have just the same chance of increasing or decreasing Z. Therefore ΔZ is given by equation 10 whether Z itself is given by equation 9 or equation 11.

Suppose Z is given by the product of measured quantities, i.e.

$$Z = AB \ldots \ldots \ldots \ldots \ldots \ldots (12)$$

Would you expect an error of ΔA on A to produce an error in Z of ΔZ such that:

$$\Delta Z = \Delta A?$$

No.

If A is increased to $A + \Delta A$, then

$$Z + \Delta Z = (A + \Delta A)B$$

$$\therefore Z + \Delta Z = AB + \Delta A \cdot B$$

$$\therefore \Delta Z = \Delta A \cdot B$$

Dividing both sides by Z

$$\frac{\Delta Z}{Z} = \frac{\Delta A \cdot B}{Z}$$

$$\therefore \frac{\Delta Z}{Z} = \frac{\Delta A \cdot B}{AB}$$

$$\therefore \frac{\Delta Z}{Z} = \frac{\Delta A}{A}.$$

Thus we see that a *fractional* increase in either A (or B) produces *the same fractional increase* in Z. Thus the fractional error, $\left(\dfrac{\Delta Z}{Z}\right)$, has contributions from $\left(\dfrac{\Delta A}{A}\right)$ and $\left(\dfrac{\Delta B}{B}\right)$. These two contributing errors may lead either to an increase or decrease in Z, and so may partially cancel each other. Therefore, this time the square root of the sum of the squares of the fractional errors is taken:

$$\frac{\Delta Z}{Z} = \left\{ \left(\frac{\Delta A}{A}\right)^2 + \left(\frac{\Delta B}{B}\right)^2 \right\}^{\frac{1}{2}} \quad \cdots\cdots\cdots (13)$$

If there are more than two factors on the right hand side of equation (12), for example, $A, B, C, D, E, \ldots$, the same rule still applies: sum the squares of all the individual fractional errors and take the square root:

$$\frac{\Delta Z}{Z} = \left\{ \left(\frac{\Delta A}{A}\right)^2 + \left(\frac{\Delta B}{B}\right)^2 + \left(\frac{\Delta C}{C}\right)^2 + \left(\frac{\Delta D}{D}\right)^2 + \left(\frac{\Delta E}{E}\right)^2 + \cdots \right\}^{\frac{1}{2}}$$

Suppose Z is expressed as a ratio:

$$Z = A/B \quad \cdots\cdots\cdots\cdots\cdots (14)$$

What effect on Z would you expect from an increase of A to $A + \Delta A$?

Increasing A to $A + \Delta A$ increases Z to $Z + \Delta Z$ so:

$$Z + \Delta Z = \frac{A + \Delta A}{B}$$

$$\therefore Z + \Delta Z = \frac{A}{B} + \frac{\Delta A}{B}$$

$$\therefore \Delta Z = \frac{\Delta A}{B}$$

$$\frac{\Delta Z}{Z} = \frac{\Delta A}{B} \cdot \frac{1}{Z}$$

$$\frac{\Delta Z}{Z} = \frac{\Delta A}{B} \cdot \frac{B}{A} \qquad \therefore \frac{\Delta Z}{Z} = \frac{\Delta A}{A}$$

Thus again, we find a fractional change in A (or B) produces a fractional change of the same magnitude in Z. Equation 13 applies to both the case where Z is calculated from equation 12 and also when it is calculated from equation 14.

To get some practice at using these formulae you might like to try questions 4 to 9, p. 44.

Finally, if Z is given by:

$$Z = KA^n \dots\dots\dots\dots\dots(15)$$

where K is a constant, then

$$\frac{\Delta Z}{Z} = n\frac{\Delta A}{A} \dots\dots\dots\dots\dots(16)$$

To help you accept this last statement let us give you an example. Suppose $A = 100$, $\Delta A = 1$, $n = 3$, and $K = 5$, then $Z = 5 \times 100^3$ which equals 5×10^6. If A were increased by ΔA, i.e., became 101, then Z would be

$$5 \times (101)^3 = 5 \times 1\,030\,301$$

$$\approx 5\,150\,000$$

Thus an error of 1 in A produces an error of 150 000 in Z. Therefore $\Delta Z = 150\,000$.

Now check that equation 16 is satisfied.

We return yet again to the experiment involving the water issuing from the tube. We now suppose the tube to be of radius, r, and length, l. The pressure difference between the two ends of the tube (providing the necessary force for pushing the water along) is P and the volume of water emerging per second is Q. Let us suppose we are interested in measuring a constant characteristic of the liquid called the coefficient of viscosity, η. The expression for η (which you are not expected to remember) is:

$$\eta = \frac{P\pi r^4}{8lQ}$$

Express the error in η in terms of the errors on the measured quantities, P, r, l, and Q.

For extra practice try questions 10 to 12, p. 45.

First of all note that r appears as the fourth power, so from equation 16 the fractional error in the factor (r^4) is given by

$$\frac{\Delta(r^4)}{(r^4)} = \frac{4\Delta r}{r}$$

Therefore, from equation (13):

$$\frac{\Delta \eta}{\eta} = \left\{ \left(\frac{\Delta P}{P}\right)^2 + \left(\frac{4\Delta r}{r}\right)^2 + \left(\frac{\Delta l}{l}\right)^2 + \left(\frac{\Delta Q}{Q}\right)^2 \right\}^{\frac{1}{2}}$$

2.6 A little common sense with errors

You have seen how the error on a final quantity is often expressed as the sum of several contributing errors. In many experimental situations, there can be gross differences in the magnitudes of these individual errors. The differences are then increased further by the practice of taking squares. For example if

$$Z = A + B$$

and $A = (100 \pm 10)\,\text{cm}$ and $B = (100 \pm 3)\,\text{cm}$, then

$$\Delta A = 10, \qquad \Delta B = 3$$

$$\Delta Z = (\Delta A^2 + \Delta B^2)^{\frac{1}{2}}$$

$$\Delta Z = (10^2 + 3^2)^{\frac{1}{2}} = (100 + 9)^{\frac{1}{2}} = (109)^{\frac{1}{2}}$$

$$\Delta Z \approx 10.4\,\text{cm}$$

As you can see even though ΔB is almost a third of ΔA, its contribution to ΔZ is almost nothing; ΔZ is nearly equal to ΔA alone. So, having shown that a contributing error is less than, say, a quarter of the dominant one, you may safely neglect it.

From such considerations, of course, it is clear that when doing an experiment it pays to find out as early as possible what the dominant errors are likely to be. You can then concentrate your time on those aspects of the experiment that are likely to reduce the dominant contributing errors.

Look at the expression you have just derived for the error on the coefficient of viscosity, η, as determined by means of the rate of flow of water through a tube. What quantity do you think should be measured with special care?

Beware of the situation where you have two rather well-measured quantities, say, $A = (102 \pm 2)$ cm and $B = (98 \pm 2)$ cm and the quantity you are after, Z, is the difference of these lengths i.e.

$$Z = A - B$$

the precision of your final result can be disappointing!

What is the percentage error on Z in this case?

Thus when measuring the difference between two almost equal quantities, these quantities must be measured with particular care.

So much for handling errors — we now turn to other aspects of practical work.

The radius of the tube, r. Its error is multiplied by a factor 4 and when the errors are compounded this becomes

$$4^2 = 16.$$

$$\Delta Z = \{(\Delta A)^2 + (\Delta B)^2\}^{\frac{1}{2}}$$
$$\text{(from equation 10)}$$
$$= (2^2 + 2^2)^{\frac{1}{2}} = (8)^{\frac{1}{2}}$$
$$\approx 2.8 \text{ cm}$$

Therefore the percentage error is

$$\frac{\Delta Z}{Z} \times 100 = \frac{2.8}{102 - 98} \times 100 \approx 70\%$$

2.7 Proportionalities

A great deal of experimental work is concerned not so much with the measurement of a particular value of some quantity, as with the study of how one quantity varies as a result of a change in some other. There are several types of such dependence.

For example the distance, x, covered in time, t, by a car travelling at constant speed, v, is given by:

$$x = vt \dots\dots\dots\dots\dots\dots(17)$$

Here x and t can take on a range of values. When two variables (x and t in this case) are related in this way we say one is *directly proportional* to the other. The constant (v) is referred to as the *constant of proportionality*.

A quantity may be directly proportional to a power or root of another quantity. In the Galileo experiment, an object released from rest travels a distance, directly proportional to the square of the time, t:

$$s = kt^2 \dots\dots\dots\dots\dots\dots(18)$$

where k is a constant. Another way of looking at the same problem is to take the square root of both sides:

$$s^{\frac{1}{2}} = (k)^{\frac{1}{2}}t$$
$$cs^{\frac{1}{2}} = t \dots\dots\dots\dots\dots\dots(19)$$

$\left[\text{where } c \equiv \left(\frac{1}{k}\right)^{\frac{1}{2}} \right]$. Thus the time of travel is directly proportional to the square root of the distance.

If a given amount of gas, maintained at a steady temperature, is subjected to an increase in pressure, then its volume decreases. Doubling the pressure halves the volume, and conversely doubling the volume halves the pressure. Thus,

$$P = K\frac{1}{V} \quad \ldots \ldots \ldots \ldots \ldots \ldots \text{(20a)}$$

where P and V are the pressure and volume respectively and the constant, K, depends on the nature of the gas, its mass and the temperature. Here we say one variable is *inversely proportional* to the other.

Proportionalities can be combined. If, for example, we have gas in a vessel of constant volume and at constant temperature, the pressure exerted is found to be proportional to the mass, M, of gas in the vessel. Therefore we can write

$$P = K'M \quad \ldots \ldots \ldots \ldots \ldots \ldots \text{(20b)}$$

(where K' is another constant not equal to K of equation 20a). But we already have that P is inversely proportional to the volume V. Thus in an experimental situation in which we allow *both* the mass and the volume of the gas to change and keep only the temperature constant, we can combine the proportionalities, i.e. combine equations 20a and 20b so:

$$P = \frac{kM}{V} \quad \ldots \ldots \ldots \ldots \ldots \ldots \text{(21)}$$

(where k is another constant).

Thus we say P is directly proportional to M, and inversely proportional to V.

Write down a formula for the amount of heat, Q, flowing through material in the form of a circular disc, given that Q is directly proportional to the difference in temperature $(T_1 - T_2)$ between the two faces; to the square of the radius, r, of the disc; to the time, t, for which the heat flows; and is inversely proportional to the thickness, d, of the disc.

$$Q = (\text{constant})\frac{(T_1 - T_2)r^2 t}{d}$$

2.8 Graphs; why use them?

If a mass is hung on the end of a wire, the length of the wire will increase — the greater the mass the greater the extension produced. The extension should be directly proportional to the applied mass up to a point called the elastic limit. Table I shows a typical set of readings.

TABLE I

Applied mass (kg)	Extension (mm)	Applied mass (kg)	Extension (mm)
5	0.2	32.5	1.7
10	0.5	35	1.8
15	0.8	37.5	1.9
20	1.0	40	2.0
22.5	1.5	42.5	2.3
25	1.3	45	2.5
27.5	1.4	47.5	2.8
30	1.5	50	3.2

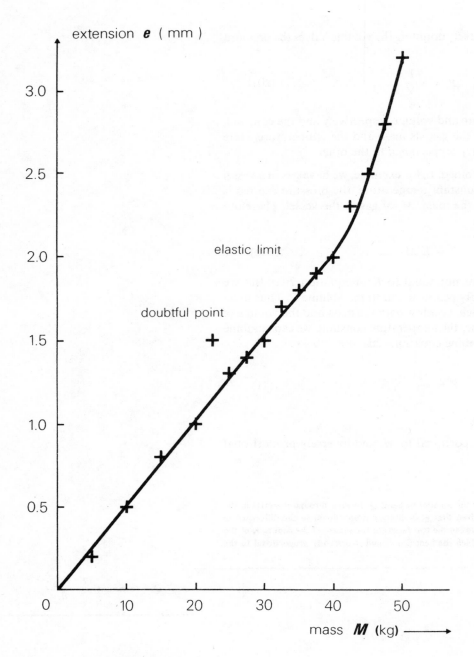

extension **e** (mm)

elastic limit

doubtful point

mass **M** (kg) ⟶

Figure 10   Extension of a loaded wire

How can you find from this set of readings the elastic limit? By just looking at the table it is not easy to answer such a question (though it is possible if you are prepared to subtract successive readings from each other and compare the differences). This is where a graph becomes useful. If the extension is plotted against the mass, a graph similar to that of Figure 10 is obtained. Here, each pair of readings in the table appears as a point on the graph located at position (e, M) where e and M are the lengths marked off along the axes to represent the appropriate values of the extension and mass respectively. The points are indicated by crosses.

It is now quite obvious where the departure from direct proportionality occurs, i.e., the point where the graph departs from a straight line.

You will see that the experimental points are scattered about the smooth curve. The amount of scatter gives at a glance a feeling for the size of the random errors involved in the experiment.

While on the subject of the scatter of the points, you will also notice that we have picked out one point with an arrow. Its distance from the line is several times the average scatter of the other points. The question is often raised in connection with such points as to whether it is permissible to ignore them. There is no easy rule we can give you, except to say that you should always be prepared to state the reason why you are ignoring a particular reading. This may be, for example, the fact that its value is the equivalent of one large division of a meter scale away from the value it ought to have — indicating a reading error. Or you may think you have wrongly recorded the reading, so as to interchange the axes for this particular point — this is especially likely to happen over ranges where the values for the two variables are rather close. In any case, you should be particularly wary about neglecting points near one of the extremes of the range of observations, because in these regions it may well be that the type of behaviour has begun to change. If possible you should, of course, repeat suspect measurements.

Another advantage of a graph is that it also makes it easy to make estimations in between measured points. For example, although no reading was taken for 17.5 kg, the graph is drawn to pass through points corresponding to all masses, including 17.5 kg. You can easily read off from the graph that the extension would have been 0.9 mm if such a mass had been used.

So what then is the function of a graph? It is a *visual aid*. The point of departure from direct proportionality, the mismeasurement, the scatter of the readings, and the interpolation *could* all have been determined from the list of readings, but you will doubtless agree that it is a lot easier to see what is going on when the results are displayed in the form of a graph. So you are advised to get into the habit of drawing a graph whenever the need arises.

2.9 Straight-line graphs; determination of the slope and intercept

In many situations it is possible to plot a graph that on theoretical grounds should yield a straight line. *Choose your variables accordingly.* For example, in the problem described by equation 17 you should plot x against t; for that described by equation 18, s against t^2; and for equation 20a, P against $\frac{1}{V}$.

The question then naturally arises as to how to draw the best straight line through the experimental points.* The determination of the slope, or gradient, of the line is also important because this represents the constant of proportionality. (In equation (17), for instance, the slope $\frac{x}{t}$ will be equal to the speed, v.)

Often it will be sufficient simply to draw in the line which appears to the eye to be about right. On other occasions however you will need to be more careful. Then you will find it convenient to use the "points-in-pairs" approach. This can be applied to a graph with points equally spaced along the horizontal axis.

---

*Ideally you would use what is called "the least squares method". Intending students of physics might like to learn about this as a black-page activity (see, for example, Squires, op. cit.).

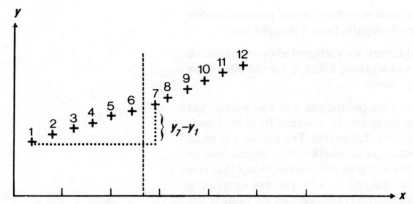

Figure 11 The point-in-pairs approach

The method is illustrated in Figure 11. First the points are divided into two equal groups, one for low values of y and the other for high values of y. The points are then paired off, one from each group so:

$$1 \text{ and } 7, 2 \text{ and } 8, \ldots 6 \text{ and } 12$$

The difference between the values of y for each pair is then found, i.e.

$$(y_7 - y_1), (y_8 - y_2) \ldots (y_{12} - y_6)$$

The mean of these differences, $\bar{D}_y$, is calculated. If the equal distance between each of these pairs of points along the x axis is D_x, e.g. $(x_7 - x_1)$, then the slope is given by

$$m = \frac{\bar{D}_y}{D_x} \dots\dots\dots\dots\dots\dots\dots(22)$$

The mean values of x and y are then determined from

$$\bar{x} = \frac{x_1 + x_2 + \cdots + x_{12}}{12} \dots\dots\dots\dots\dots(23)$$

and

$$\bar{y} = \frac{y_1 + y_2 + \cdots + y_{12}}{12} \dots\dots\dots\dots\dots(24)$$

The best line by this method is then the one of slope, m, passing through the point $(\bar{x}, \bar{y})$.

How would you suggest we estimate an error for m?

The six differences $(y_7 - y_1)$ etc. provide six measurements of quantity $\bar{D}_y$, and hence six measurements of m. They can be used therefore in the normal way for estimating errors (see the section 2.2 on random errors, pp. 7ff.).

But, you may ask, why stop at six measurements of m, why not take eleven by evaluating $(y_2 - y_1)$, $(y_3 - y_2) \ldots (y_{12} - \dot{y}_{11})$ and so improve still further the precision? Beware of the trap — can you see it?

If you were to evaluate $\bar{D}_y$ from these eleven readings you would get

$$\frac{(y_{12} - y_{11}) + (y_{11} - y_{10}) + (y_{10} - y_9) \cdots + (y_2 - y_1)}{11} = \frac{y_{12} - y_1}{11}$$

All the readings from y_{11} to y_2 inclusive have disappeared — you might just as well not have taken them! This disaster is avoided if you divide the readings into two main groups and pair off points one from each group, as described.

For practice try question 13, p. 45.

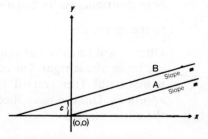

Figure 12 Slope and intercept of a straight line graph

If for some reason it were not possible to have the readings spaced at exactly equal intervals along the horizontal axis, how could you estimate m by a method closely similar to that of the points-in-pairs method?

You would evaluate six separate values of m,

$$\frac{(y_7 - y_1)}{(x_7 - x_1)}, \quad \ldots \quad \frac{(y_{12} - y_6)}{(x_{12} - x_6)}, \text{ and average}$$

If, as above, the points are not exactly equally spaced along the x axis, why is it still important that they should be roughly equal?

If they were not roughly equal, the individual measurements of m based on the wider spaced points would give better estimates than those based on the closer points. Remember the various formulae you have been given depend on individual readings having the same reliability.

If a variable y is directly proportional to another variable x, then there exists a theoretical relationship between them: $y = mx$, where m is the constant of proportionality. The pairs of (y, x) measurements should be on a line of slope m passing through the origin, i.e., the point $(0, 0)$. Such a line is shown as line A in Figure 12. (Equation 17 is an example of a directly proportional relation. It is obvious that the distance travelled in zero time must be zero, thus a graph of x against t ought to pass through the point $(x = 0, t = 0)$). However, the points-in-pairs procedure gives a line that, because of experimental imprecision, does not necessarily pass exactly through the origin. What should you do in these circumstances? If you know on theoretical grounds that the line should go through the origin, do you constrain it to do so? Some people think that this is the best procedure; but our advice to you is that you should not. We wish you to use the points-in-pairs method as usual. Our reason is simply that the amount by which the line misses the origin may contain information about a possible systematic zero error in the apparatus. If such an error is present, and you force the line to go through the origin, you will reduce the accuracy of your determination of the gradient.

Proportionality is not the only kind of relationship that can give a straight-line graph. There are others that give a straight line not passing through the origin. Line B in Figure 12 is an example of one. It has the same slope, m, as line A but cuts the y and x axes at points other than the point $(0, 0)$. The distance from the origin to the point where the line or curve cuts an axis is called the *intercept* on that axis. Points lying on such a line will have pairs of y and x measurements related by the equation:

$$y = mx + c \ldots\ldots\ldots\ldots\ldots\ldots\ldots(25)$$

Note that the form of this equation is such that when $x = 0$, y is no longer 0 but has the value c; c is then the intercept on the y axis. Equation (25) is the most general form of relationship between two variables, y and x, such that a graph of one against the other gives a straight line. The directly proportional relation, $y = mx$, is one particular type having $c = 0$.

If c is the intercept on the y axis, what is the intercept on the x axis?

When the line cuts the x axis, $y = 0$
$\therefore 0 = mx + c \quad \therefore mx = -c$
Thus the intercept on the x axis is $-c/m$.

How are the intercepts determined? If the plotted graph shows the true origin of the axes, i.e., the point $(0, 0)$ as in Figure 12, then there is no problem: you draw the best line through the points, extend it to strike the axes, and read off the distances. If however, the measurements are bunched together over a small range a considerable distance from the origin, as in

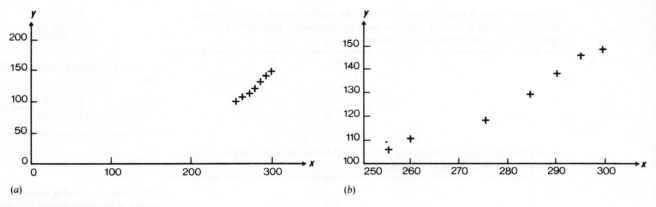

(a)   (b)

Figure 13   Choice of scales

Figure 13a, it is preferable to spread the points out over the page, as in Figure 13b, at the cost of not showing the true origin (the point $(y = 100, x = 250)$ is called a 'displaced origin'). The problem is how to estimate the intercepts on the true axes when these are not shown. This is not difficult: having estimated m from the points-in-pairs approach, and knowing that the best line passes through the point $(\bar{x}, \bar{y})$, you have a simple equation for this point:

$$\bar{y} = m\bar{x} + c$$

$$\therefore \ c = \bar{y} - m\bar{x} \ldots\ldots\ldots\ldots\ldots\ldots (26)$$

This gives the value of the intercept on the y axis, and $-c/m$ gives the corresponding intercept on the x axis.

For practice try question 14, p. 45

Before leaving the subject of straight-line graphs, let us draw your attention to the advantages of using *logarithmic graph paper* in certain types of investigations.

Suppose there are two variables, y and x, related by the expression

$$y = K \cdot k^x \ldots\ldots\ldots\ldots\ldots\ldots (27)$$

where K and k are constants. One such expression is given in section 5.4 of Unit 2:

$$N_t = N(\tfrac{1}{2})^{t/T\frac{1}{2}}$$

In this example y represents the number of radioactive nuclei left after time t (N_t), K is the original number of nuclei (N), k is a constant ($\frac{1}{2}$), and x is the time during which the decays take place expressed in terms of half-lives ($t/T_{\frac{1}{2}}$).

Equation 27 appears to have little in common with the general expression for a straight line: $y = mx + c$. Indeed a plot of y against x does not give a straight line as can be seen from Figure 36 of Unit 2. However, we can express equation 27 in a somewhat different form by taking logarithms of both sides.*

$$\log y = x \log k + \log K$$

$$\log y = mx + c \ldots\ldots\ldots\ldots\ldots\ldots (28)$$

where m and c are constants ($m \equiv \log k$ and $c \equiv \log K$). Thus, if x is plotted against $\log y$ instead of y, a straight line will result. Because it is

Refer to the MAFS if you are unfamiliar with logarithms.

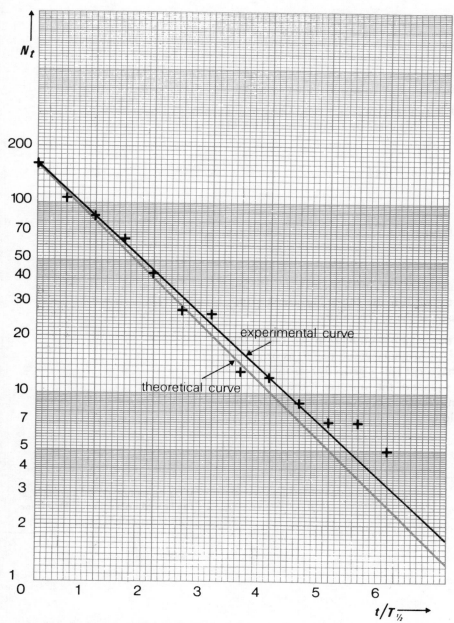

Figure 14   Use of logarithmic graph paper

much easier to judge whether points lie on a straight line than on a curve of a particular shape, you ought in this problem to plot log *y* against *x*.

Now there is nothing to stop you looking up the logarithms of all the *y* readings and plotting the graph on ordinary graph paper. However, it is more convenient to use a graph paper that does the job for you. The paper used in Figure 14 is called "log × 10ths". The horizontal scale is an ordinary one in which the large divisions are divided into tenths and each division has the same size. The vertical scale is the log scale and, as you see, its divisions become progressively compressed towards the upper end (in the same way as the logarithms of numbers increase more slowly than the numbers themselves). If you plot *y* on this log scale, the points become distributed over the graph paper in the same way as if you had plotted log *y* on ordinary graph paper.

To illustrate the advantage of using this type of paper we show in Figure 14 a plot of the same data as was used in Figure 36 of Unit 2. Instead of the points lying about a curve, they are now seen to follow a straight line.

This is not the only type of logarithmic graph paper. To demonstrate the usefulness of another type, let us suppose we have an expression of the form.

$$y = k \cdot x^m \quad \ldots\ldots\ldots\ldots\ldots\ldots(29)$$

(Equation 18 is an example of such an expression: $s = kt^2$.) Once again this is not in the form: $y = mx + c$. However, by taking logarithms of both sides, we get

$$\log y = m \log x + \log k$$

or

$$\log y = m \log x + c \quad \ldots\ldots\ldots\ldots\ldots(30)$$

(where $c \equiv \log k$).

Thus a plot of $\log y$ against $\log x$, instead of y against x, yields a straight line. In order to plot this conveniently we can use log × log graph paper, in which both scales are compressed at the upper values.

But why go to this trouble in the example quoted? If we know that $s = kt^2$, why not plot s against t^2 on ordinary graph paper, as originally suggested, and get a straight line that way? This is perfectly reasonable — *if* you know that s depends on the square of the time, i.e., if the value of m in equation (29) is known. The bonus that comes with plotting the variables on a log × log paper is that it allows you to determine m.

How can this be done?

If $\log y$ is plotted against $\log x$ then, as can be seen from equation 30, m is the slope of the straight line.

2.10 Miscellaneous hints on how to plot graphs

(a) In an experiment it is usual to change one quantity (the *independent variable*) by given amounts and see what effect is produced on the other quantity (the *dependent variable*). It is usual to plot the independent variable along the horizontal axis (called the *abscissa*) and the dependent variable along the vertical axis (called the *ordinate*).

(b) Choose scales such that the points are as widely spread over the sheet of paper as possible. We have already indicated that Figure 13b is preferable to Figure 13a.

(c) Choose simple scale divisions so as to reduce the labour involved in estimating to fractions of a division (and also to reduce the risk of making a mistake). It is recommended that each small division should be 1, 2 or 5 times some power of 10. For example 0.1, 0.2, 0.5, 1, 2, 5, 10, 20, 50, etc.

(d) To avoid confusion with specks of dirt, plot points as crosses (+) or circles (⊙), and not as dots.

(e) Always label the axes with the name of the quantity being plotted along that axis and also the units in which the quantity is measured.

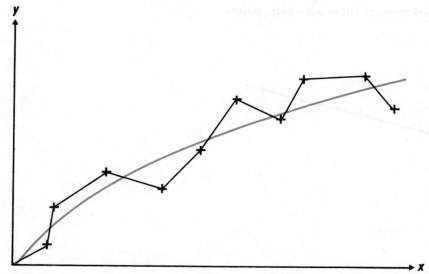

Figure 15

(f) When drawing the "best curve" through the points do not join up the individual points as shown by the black line in Figure 15, but put a smooth curve through them, as is shown by the red line. The slowly-changing variation of the latter is more likely to correspond to reality. There are exceptions however as you will see later (Unit 6).

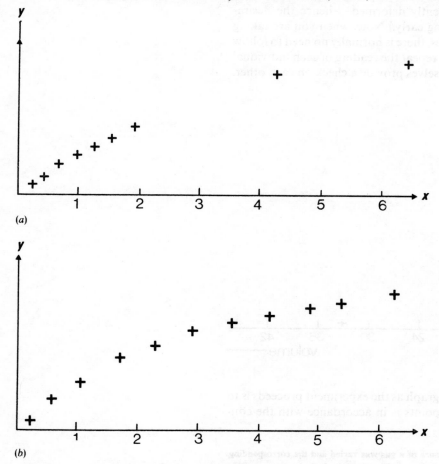

(a)

(b)

Figure 16   (a) Readings unevenly spread; (b) Readings evenly spread

(g) When taking readings generally spread them out evenly over the range of values of the quantity being measured. Figure 16a is poor, 16b is better. An exception to this rule is when you want to measure an intercept, as in Figure 17. Then it is desirable to have a few extra points close to the axis.

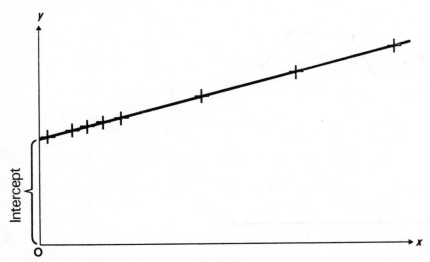

Figure 17   Extra points near the axis to measure an intercept

(h) Plot a graph as your experiment proceeds. In this way you can check immediately if a point is so widely off as to need repeating. (In Figure 10 you *could not* have gone back after the experiment was finished to check the point out of line because once a specimen has exceeded its elastic limit it becomes permanently deformed — hence the desirability of spotting this kind of thing early.) Note, when you are taking a series of readings of two variables, there is normally no need to follow the instruction of Section 2.1 and repeat the reading of each individual setting; the various settings themselves provide a check on each other.

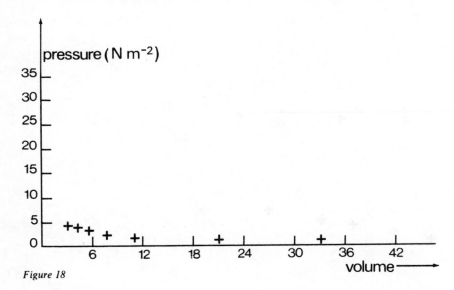

Figure 18

A further reason for plotting the graph as the experiment proceeds is to see whether the distribution of points is in accordance with the considerations outlined under (g).

In an experiment in which the pressure of a gas was varied and the corresponding volume measured, the results shown in Figure 18 were obtained. On how many counts can you fault the plotting of the points?

You may have mentioned some of the following faults:
(1) The independent variable (the pressure) should have been plotted along

2.11 Planning an experiment

The time available for performing an experiment is nearly always limited. Practical work is only a part of your weekly assignment. You will be working under constant pressure to do a "reasonably good" job and then move on as quickly as possible to something else. This is not unlike the situation in which the professional researcher finds himself — being harried by his employer, or by the thought that a rival may come up with the answer sooner than he does. The need to use time efficiently makes it essential to *plan* an experiment.

You have been told to repeat readings. But how can you know in advance how many readings you can afford to take of a given quantity? The answer obviously depends on the time each reading takes. You have seen that the way in which readings are spread over the available range affects the quality of the final result. But how can you know what the appropriate interval between readings is before you know how large the range is and how long it takes to make each measurement? You have been told to concentrate on those aspects of the experiment that give rise to the largest contributing errors. But how can you tell in advance which these are?

These questions can usually be answered by a preliminary experiment. It nearly always pays to have a preliminary run through before embarking on the experiment proper — in short, a rehearsal.

This way you can check that everything is functioning correctly, and you can make a provisional estimate of the size of the various errors and so decide how many readings to take of each quantity.

2.12 Keeping a laboratory notebook

Record your measurements in a notebook. All comments, observations, calculations, etc. should go into it. It is bad practice to record anything on loose scraps of paper — they only get lost. Make a note of the date on which you take the measurements. When taking a reading, do not do a little mental arithmetic (like subtracting a zero error) and record only the end product. Record *direct* observations and in so doing avoid making mistakes that cannot later be checked ("that zero error — did I *add* it when I should have *subtracted* it?").

Readings should be recorded in tables, each column appropriately headed and the units specified. If the units are stated at the top of a column it is unnecessary to repeat them for each new reading.

Important quantities should be underlined to make them stand out boldly.

If you write a number incorrectly, do not overwrite it — there is always a danger of wrongly deciphering the overwritten number. Always cross out the wrong entry and record the correct number alongside it.

Do not be afraid to spread your account over several pages. The more paper you use the sooner you will have to buy another book, but saving paper in a laboratory notebook is a bad form of economy. It is never a waste of paper or of your time to jot down all manner of notes about the way the experiment is progressing — for example, the reason you have decided to repeat some readings, a note of anything you have noticed out of the ordinary, suspicions that some part of the equipment may not be functioning correctly, checks you have made, etc. The temptation to skimp on these comments and simply dash down a set of readings is so strong you will doubtless succumb — but for sure you will regret these lapses. It is quite remarkable how a few cryptic numbers so self-explanatory when written, rapidly become utterly meaningless!

Bear in mind that in keeping a notebook you are following the practice of the professional research worker. His experiments extend over months

the abscissa.
(2) The units for volume are not labelled.
(3) The range of the scale for pressure is poorly chosen — the range should have been 0–5 Nm^{-2} instead of 0–35 Nm^{-2}. In this way the points would have been better distributed over the page.
(4) The subdivisions along the abscissa are awkward to estimate.
(5) The plot should not have been P against V anyway — it should have been P against $1/V$ in order to get a straight line.

or years before the final result emerges; under these circumstances, the making of copious notes is clearly essential for continuity.

2.13 Hints on performing calculations

Most calculations can be made to sufficient accuracy with a good 10-in. slide rule engraved with the basic A, B, C, and D scales with a reciprocal scale in red down the centre. You would be well advised to buy one. Instructions on how to use a slide rule are usually supplied by the manufacturer. The accuracy attainable with such an instrument is about 0.2 per cent. If this is not thought good enough (when for example you are computing two nearly equal numbers that are subsequently to be subtracted) you will have to resort to logarithms.

Often you will take a reading that has then to be put into a formula and multiplied by various factors to produce a final answer. If you are making several determinations of this quantity with a view to averaging them, average the initial readings and not the final results. This way you will have to multiply only once and so will avoid unnecessary labour.

Another labour-saving tip is to use a "false mean". Suppose for instance you had to find the mean of the following numbers:

$$1372.4, \ 1373.5, \ 1378.7, \ 1371.9, \ 1377.8, \ 1381.2, \ 1376.9$$

Would you add them as they are and divide by 7? It is quicker to adopt a convenient "false mean", in this case for example 1370.0 and subtract it from each number in turn and average the differences:

$$\frac{2.4 + 3.5 + 8.7 + 1.9 + 7.8 + 11.2 + 6.9}{7} \approx 6.1$$

The required mean is then the sum of this averaged difference and the false mean:

$$6.1 + 1370.0 = 1376.1$$

Get into the habit of checking calculations — It is so easy to make an arithmetic slip. When checking, do not carry out the calculation in exactly the same way as you did the first time, otherwise you are likely to make the same mistake again. Take, for example, the expression:

$$K = \frac{2\pi \times 0.638 \times (27.1)^2 \times 1.28}{96.4} = 39.2$$

The first time you could evaluate it in the order in which the numbers appear; the second time in reverse, i.e.,

$$1.28 \times 27.1 \times 27.1 \times 0.638 \times \pi \times 2/96.4$$

Where does the decimal point go? In deciding this, round the numbers off so:

$$K \approx \frac{2 \times 3 \times (6 \times 10^{-1}) \times (30)^2 \times 1}{100} \approx \frac{36 \times 10^{-1} \times 9 \times 10^2}{10^2} \approx 30$$

As a final check, always ask yourself whether your result is *reasonable*. Sometimes there is no way of telling whether a result is reasonable, but it is surprising how often mistakes can be detected by simply asking oneself this simple question. (Did you *really* have 96.75 m³ of water in that 200 cm³ beaker? Does that "best straight line" on the graph *look* as though it goes through the points?)

Finally a word about *significant figures*. As you have seen, all measurements are subject to inaccuracy. This means you can never quote an exact value for a measured physical quantity. If the mass of an object were estimated to lie somewhere between 3.75 and 3.85 g, then the result would be quoted as 3.8 g. There would be no point in adding any figures to the right of the "8" as they are liable to be wrong. Such a measurement is said to have been made to two significant figures.

The same result could have been expressed in terms of kilogrammes, *viz.*, 0.0038 kg. The noughts to the right of the decimal point are necessary in order to establish the position of the decimal point relative to the digits "38". The lack of figures to the right of the "8" still indicates that the accuracy of the determination is to two significant figures.

Suppose the same mass were expressed in milligrammes: 3800 mg. Once again the noughts are necessary in order to place the digits "38" relative to the decimal point (i.e. to indicate that we are now dealing with 3 thousand, 8 hundred). But now an element of uncertainty has crept in — are those noughts there *only* to indicate the position of the decimal point, or do they also mean that the measurement has been performed to an accuracy of four significant figures, i.e. the mass has been measured to lie between 3799.5 mg and 3800.5 mg? There is no way of telling unless an error is specifically quoted along with the measured value — for example (3800 ± 50) mg. It is best to avoid the confusion by using an appropriate power of 10: 3.8×10^3 mg for a determination to two significant figures; 3.800×10^3 mg to indicate four significant figures.

Always be on your guard against writing down a long string of numbers that have no significance — it wastes your time and misleads others. Meaningless digits are particularly liable to arise from calculations. Suppose for example it takes you 23 seconds to walk out of your house to your car, about 1 hour to complete the journey by road to your work, and 2 minutes to walk from the car park to your office at the other end — what is the total journey time?

1 hr 2 min 23 s?

Probably not. To quote such a result, you would have to time the car journey and the walk at the other end to an accuracy of 1 second — there is nothing to indicate that such care was exercised when timing the last two portions of the journey. The best estimate you could make of the total time would be "about an hour", i.e. the accuracy of your final result is governed by the accuracy of your worst measured quantity.

Given that the lengths of 4 sticks are 0.46 cm, 27.6 cm, 3 cm and 0.12 cm, what is the total length of the sticks when placed in a straight line end to end?

Multiplication can quickly lead to strings of meaningless digits. Suppose you wished to find the volume of a box whose sides are measured to be 1.4 cm, 27.1 cm and 103.2 cm:

$$1.4 \times 27.1 \times 103.2 = 3915.408 \text{ cm}^3$$

How many of these digits are meaningful? Note that the measurement of the first length has only two significant figures — you would not be surprised if the real value were 1.45 or 1.35 cm. If it were 1.45 cm the volume would be 4055.244 cm³, and if 1.35 cm the volume becomes 3775.572 cm³. Thus two significant figures at most can be claimed for the final result, i.e. the volume is 3900 cm³.

Is that a good way of expressing the final result?

So the accuracy of a multiplication (or division) can be no better than that of the least accurate quantity appearing as a factor.

(Some formulae involve factors that are pure numbers. For example "2" and "π" in the formula $2\pi r$ relating the circumference of a circle to its radius, r. For such numbers there is of course no inaccuracy. Thus the

---

The sum of the lengths is:

	0.46
	27.6
	3.
	0.12
Total =	31.18

But the length of the third stick is known only to the nearest centimetre so the sum ≈ 31 cm

---

No. It should be expressed as 3.9×10^3 cm³ (or 39×10^2 cm³) to avoid the two noughts being interpreted as significant figures.

single digit "2" implies "2.00000000..." and does not restrict the number of significant figures of the final result).

2.14 The writing of final reports

Through his experiments Henry Cavendish, the eighteenth-century British physicist, came to a deep understanding of electrical phenomena. He had however one serious fault — he rarely took the trouble to write up his findings. The result was that for many years people continued working on problems he had long ago solved. For example, he discovered Ohm's Law of electrical resistance fifty years before Ohm did. The law is quite rightly ascribed to Ohm though, because it was the German schoolmaster who made the information available to others. Moral: you may get personal satisfaction from doing good experimental work but, if you wish to get credit for it and to benefit the community at large, you have to develop an ability to communicate your findings to others. Moreover you must do this in sufficient detail so that others may decide for themselves whether to believe your result or not. This is the purpose of asking you to submit to your tutor (from time to time) a final report on an experiment. This report is not to be confused with the account you always keep in your own laboratory notebook — that is your own private diary. The final report is your showpiece — it is on this that you are judged.

Perhaps you have already had some experience of writing up practical work from your schooldays. If so, we ask you to set aside some of the conventions you were originally trained to observe. For example, do not begin with a dreary list of the apparatus you have used. Neither should you tell us how *we* ought to do the experiment — avoid the imperative ("Place the thermometer in the test tube. Stir the contents continuously. Connect up the apparatus as in the diagram."). You could use instead the passive voice, past tense ("The thermometer was placed in the test tube. The contents were stirred continuously. The apparatus was connected up as shown in the diagram."). It is quite all right to use the occasional "I" ("I decided to check whether any leads might have come loose.").

We want you to aim at producing a report in the style of a paper to a scientific journal. An account modelled on an industrial report would perhaps be of more relevance to some of our students but, if for no other reason than to ease the burden on our tutors, we would like a uniform style of presentation to be adopted. The following points should therefore be noted:

(a) Title

Scientists are busy people and when consulting the latest journal their first action is to run their eye rapidly through the titles in the index. If you want to catch the attention of a reader who might be interested in your work the title must communicate. It should carry as much information as possible without being too unwieldy. If you were writing up the TV experiment of Unit 2 for example, a title such as "An Experiment with a Cathode-ray Oscilloscope" would be too vague. "A Measurement of the Lifetime of Positively Charged Cosmic Ray Muons stopped in Lead using Scintillation Counters and a Cathode-ray Oscilloscope" would be too long. "A Measurement of the Lifetime of Muons using a Cathode-Ray Oscilloscope" would be acceptable.

(b) Date

This is an important item in many a research paper ('Who discovered it first?'). In your case the date is not very crucial, but for all that, it is still worth including as you may wish to look back and see how your work has progressed with time.

(c) Abstract

This is a summary of the experiment, featuring particularly the result and principal conclusions and should *always* be included in an account. It should be 20 to 100 words long. This restriction makes the writing of abstracts a difficult art. It is a skill, however, you must try to master; many readers of a science journal get no further than the abstract. If they are not actually working in your line, they do not have the time to read the whole paper – they merely wish to keep in touch with general trends in your field. Similarly, if you work in industry, the directors of the firm may only have time to read the abstract, the detailed study of the whole report being delegated to your immediate supervisor.

(d) Main account

The form of the main account will vary from experiment to experiment. It should include various sub-headings such as "Principles of the Method", "Procedure", "Results", etc. You should *always* begin it, however, with "Introduction".

The purpose of the introduction is to set the scene in general terms. In it you describe the purpose of the experiment and give a little theoretical background to the problem. If the aim of the experiment is to verify a formula, it should be quoted but not *derived* – an experiment is not a theoretical exercise. You should avoid all details and technicalities in the introduction.

You may wonder at what level you should aim this introduction – who is it intended for? When writing the opening section of a scientific paper, a researcher has in mind a fellow scientist who, though not working in the field, has had his interest sufficiently stimulated by the abstract to read on a little further. The writer knows that a plunge into technicalities at this stage will put off potential readers — hence the gentle approach.

We suggest you adopt a similar attitude and also have in mind a reader who is your equal — a fellow student, one of average ability pursuing the same course as you except that he has for some reason or other missed out on the last Unit or two — hence his need for some background information.

When describing apparatus bear in mind that a well-labelled diagram can say more than many paragraphs of prose.

Your account may contain tables, or graphs, or drawings of specimens. Make a point of referring to them somewhere in the text. ("The readings are set out in Table I." "W is plotted against x in Figure 2.") It can be annoying and confusing for a reader to come across tables and graphs scattered haphazardly throughout the account with no specific reference to them.

When drawing specimens, you should use a pencil rather than a pen so as to be able to make corrections if necessary. The drawings should be made very large; do not think that by using a thick pencil and cramming the drawing into a small space you can avoid being careful over details — if your tutor cannot clearly see the details he has no option but to assume that they are not there!

Diagrams and graphs should all be referred to as "figures" and should be numbered consecutively. For example, Figure 1 might be a diagram of the apparatus, Figure 2 a graph, etc.

(e) Conclusions and discussion

This the most important section of all — it is from this section that your tutor will judge whether you are developing critical awareness. Here are some guide lines as to what you might like to include:

a discussion of assumptions and approximations made;

a discussion of the consistency of the readings (do repeated readings give values that agree reasonably with each other?);

a discussion of how you came to assign your quoted error—how much of it was attributable to random effects, how much to systematic errors;

what were the limitations of the apparatus? If you were to redesign the experiment to give a better result, what features would you regard as being in most need of improvement?

a discussion of any unexpected behaviour;

a comparison of your result with theory or with some accepted value of the quantity being measured;

a discussion of the significance or relevance of the results;

suggestions as to how the phenomenon being studied could be investigated more thoroughly and extensively.

Finally let us stress that your account should be a full and faithful record of what you have done. If, for example, during the experiment, you took certain precautions and made certain checks, but subsequently omitted to point these out in your account, your tutor can only assume that you did not do them, and moreover that it did not even occur to you that you ought to have done them. *You cannot expect to get credit for anything not described in your account.*

2.15 Summary: a 50-point check list

For your convenience we here summarize the various hints on how to do practical work contained in this section. Keep referring to them throughout the year until with experience they become second nature to you.

Averaging

1. Measurements should be repeated where possible.
2. The mean value, $\bar{x}$, of readings $x_1, x_2, \ldots x_n$ is:

$$\bar{x} = \frac{x_1 + x_2 + \cdots + x_n}{n} \quad \ldots\ldots\ldots\ldots\ldots (1)$$

Random errors

3. The spread of readings about the mean value is characterized by the standard deviation of the sample, s:

$$s \approx \frac{5}{4} \frac{|d_1| + |d_2| + \cdots + |d_n|}{n} \quad \ldots\ldots\ldots\ldots (5)$$

where

$$|d_n| \equiv |x_n - \bar{x}| \quad \ldots\ldots\ldots\ldots\ldots (2)$$

68 per cent of the readings lie within $\pm s$ of $\bar{x}$.

4. An estimate of the standard error on the mean of n readings is given by

$$s_m \approx \frac{s}{(n-1)^{\frac{1}{2}}} \quad \ldots\ldots\ldots\ldots\ldots (4)$$

Systematic errors

5. Keep them as small as possible (note for example parallax, zero error, stop-watch calibration, etc.).

Combining random and systematic errors

6. To find the resultant error, E, of individual errors, $e_1, e_2, \ldots e_n$, sum the squares and take the square-root:

$$E = (e_1^2 + e_2^2 + \cdots + e_n^2)^{\frac{1}{2}} \quad \ldots\ldots\ldots\ldots (8)$$

Errors on sums, differences, etc.

7. When $Z = A + B$ and the errors on A and B are ΔA and ΔB, then the error, ΔZ, on Z is:

$$\Delta Z = [(\Delta A)^2 + (\Delta B)^2]^{\frac{1}{2}} \quad \ldots\ldots\ldots\ldots\ldots(10)$$

8. When $Z = A - B$ then

$$\Delta Z = [(\Delta A)^2 + (\Delta B)^2]^{\frac{1}{2}} \quad \ldots\ldots\ldots\ldots\ldots(10)$$

9. When $Z = AB$:

$$\frac{\Delta Z}{Z} = \left[\left(\frac{\Delta A}{A}\right)^2 + \left(\frac{\Delta B}{B}\right)^2\right]^{\frac{1}{2}} \quad \ldots\ldots\ldots\ldots(13)$$

10. When $Z = \dfrac{A}{B}$:

$$\frac{\Delta Z}{Z} = \left[\left(\frac{\Delta A}{A}\right)^2 + \left(\frac{\Delta B}{B}\right)^2\right]^{\frac{1}{2}} \quad \ldots\ldots\ldots\ldots(13)$$

11. When $Z = KA^n$:

$$\frac{\Delta Z}{Z} = n\frac{\Delta A}{A} \quad \ldots\ldots\ldots\ldots\ldots\ldots(16)$$

A little common sense with errors

12. Any contributing error less than a quarter of the dominant error can be ignored.
13. Special care is needed when two almost equal quantities have to be subtracted.

Straight line graphs

14. Whenever possible choose your variables so as to get a straight line. This applies to variables related by an expression of the form $y = mx + c$.
15. To find the "best straight line" by the points-in-pairs approach divide the points into two groups. If there were 12 points for example, you would find the mean, $\bar{D}_y$, of the differences $(y_7 - y_1) \ldots (y_{12} - y_6)$. Then if D_x is the equal spacing of these pairs of points along the x axis, the slope, m, is given by

$$m = \bar{D}_y/D_x \quad \ldots\ldots\ldots\ldots\ldots(22)$$

and the line passes through the points $(\bar{x}, \bar{y})$ where

$$\bar{x} = \frac{x_1 + x_2 + \cdots + x_{12}}{12} \quad \ldots\ldots\ldots\ldots(23)$$

and

$$\bar{y} = \frac{y_1 + y_2 + \cdots + y_{12}}{12} \quad \ldots\ldots\ldots\ldots(24)$$

If the spacing along the x axis is not constant, then average the individual slopes:

$$\frac{y_7 - y_1}{x_7 - x_1}, \quad \ldots \frac{y_{12} - y_6}{x_{12} - x_6}$$

16. If a line has slope m and passes through the point $(\bar{x}, \bar{y})$, its intercept on the y axis is given by $c = \bar{y} - m\bar{x}$ and on the x axis by $-c/m$.

Miscellaneous hints on how to plot graphs

17. Plot the dependent variable (i.e. the one that changes as a result of alterations you make to the other) along the vertical axis (ordinate) and the independent one along the horizontal axis (abscissa).

18. Choose scales so the points are widely spread over the page.
19. Use simple scale divisions.
20. Plot points as crosses or circles.
21. Label the axes with the *name* of the quantity and the *unit*.
22. Put a smooth curve through the points – not a zig-zag.
23. Spread readings evenly over the range unless an intercept on an axis is to be measured (in which case have several readings close to the axis concerned).
24. Plot the graph as the experiment proceeds.

Planning an experiment

25. If possible perform a quick preliminary experiment as a rehearsal.

Keeping a laboratory notebook

26. Do not use scraps of paper – put readings and comments into a special notebook kept for the purpose.
27. Record *direct* observations.
28. Record readings in tables with columns headed with the name and unit.
29. Underline important quantities (for example mean values).
30. Do not overwrite incorrect numbers – cross out the wrong number and record the new one alongside. .
31. Write copious notes on all you do.

Hints on performing calculations

32. Use a good 10-in slide-rule (or logarithms).
33. Average raw readings rather than processed data in order to save the labour of unnecessary multiplications.
34. Check calculations in reverse order of multiplication.
35. Round off numbers when deciding on the position of the decimal point.
36. Ask yourself whether the result of a calculation or measurement looks reasonable.
37. Use powers of 10 to avoid giving the impression that noughts used for placing the decimal point are significant figures.
38. Suppress meaningless digits arising from calculations. The final result of a multiplication or division can have no more significant figures than those possessed by the factor with the fewest.

The writing of final reports

39. Write mainly in the passive voice, past tense. The occasional use of "I" is acceptable.
40. Avoid a title that is either too short and vague or too long-winded and clumsy.
41. Include the date.
42. Always write an abstract of 20 to 100 words summarizing the result and main conclusions.
43. Always write a general non-technical introduction to set the scene. It should be aimed at the level of a fellow student.
44. Do not give theoretical derivations of formulae.
45. Draw well-labelled diagrams of apparatus.
46. Refer in the text to all tables and figures.
47. Drawings of specimens should be done initially at least in pencil, and should be large. You should indicate the scale.
48. Always write a section entitled "Conclusions". In it discuss assumptions, approximations, consistency of readings, random and systematic errors, limitations of apparatus, suggestions for improvements, abnormal behaviour, comparison of result with that expected, etc.
49. Every physical quantity calculated should have a unit, the correct number of significant figures, and an estimated error.
50. Mention all precautions and checks – you cannot get credit for them otherwise.

Section 3

Units

3.1 SI units

The system of units adopted by the Open University is known as SI. This is an abbreviation for Système International d'Unités. It was formally approved in 1960 by the General Conference of Weights and Measures and the Royal Society is actively encouraging its use in the United Kingdom. It is an expanded version of the MKSA system.

3.1.1 Basic units

There are six basic units:

Physical quantity	Name of unit	Symbol for unit
length	metre	m
mass	kilogramme	kg
time	second	s
electric current	ampere	A
thermodynamic temperature	kelvin	K
luminous intensity	candela	cd

Symbols for units do not take a plural form (e.g. 5 kilogrammes is written 5 kg not 5 kgs).

3.1.2 Supplementary units

In addition to the six basic units there are two supplementary units associated with angular measurements. They are dimensionless.

Physical quantity	Name of unit	Symbol for unit
plane angle	radian	rad
solid angle	steradian	sr

3.1.3 Derived SI units

The following units are derived from the basic units, but for convenience are assigned special names:

Physical quantity	Name of unit	Symbol for unit	Definition of unit
energy	joule	J	$kg\,m^2\,s^{-2}$
force	newton	N	$kg\,m\,s^{-2} = J\,m^{-1}$
power	watt	W	$kg\,m^2\,s^{-3} = J\,s^{-1}$
electric charge	coulomb	C	$A\,s$
electric potential difference	volt	V	$kg\,m^2\,s^{-3}\,A^{-1} = J\,A^{-1}\,s^{-1}$
electric resistance	ohm	Ω	$kg\,m^2\,s^{-3}\,A^{-2} = V\,A^{-1}$
electric capacitance	farad	F	$A^2\,s^4\,kg^{-1}\,m^{-2} = A\,s\,V^{-1}$
magnetic flux	weber	Wb	$kg\,m^2\,s^{-2}\,A^{-1} = V\,s$
inductance	henry	H	$kg\,m^2\,s^{-2}\,A^{-2} = V\,s\,A^{-1}$
magnetic flux density	tesla	T	$kg\,s^{-2}\,A^{-1} = V\,s\,m^{-2}$
luminous flux	lumen	lm	$cd\,sr$
illumination	lux	lx	$cd\,sr\,m^{-2} = lm\,m^{-2}$
frequency	hertz	Hz	s^{-1}
customary temperature, θ_c	degree Celsius	°C	$°C = K - 273.15$

3.1.4 Units to be allowed in conjunction with SI

Certain units have proved so popular among scientists that, although strictly speaking there is no further need for such units, they are nevertheless being retained:

Physical quantity	Name of unit	Symbol for unit	Definition of unit
length	parsec	pc	$30.87 \times 10^{15}\,m$
	centimetre	cm	$10^{-2}\,m$
area	barn	b	$10^{-28}\,m^2$
	hectare	ha	$10^4\,m^2$
volume	litre	l	$10^{-3}\,m^3$
pressure	bar	bar	$10^5\,N\,m^{-2}$
mass	tonne	t	$10^3\,kg = 1\,Mg$
kinematic viscosity, diffusion coefficient	stokes	St	$10^{-4}\,m^2\,s^{-1}$
dynamic viscosity	poise	P	$10^{-1}\,kg\,m^{-1}\,s^{-1}$
magnetic flux density (magnetic induction)	gauss	G	$10^{-4}\,T$
radioactivity	curie	Ci	$37 \times 10^9\,s^{-1}$
energy	electron-volt	eV	$\approx 1.6021 \times 10^{-19}\,J$

The common units of time (e.g., hour, year) will persist, and also the angular degree. Also in the Open University we shall retain the ångström (Å) $(10^{-10}\,m)$ as a unit of length because of its convenience when referring to the size of an atom.

3.1.5 Fractions and multiples

In some contexts our units may be too large or too small. We can of course always get over the problem of writing many zeros before and after the decimal point by using powers of 10, for example, 10^{-6} kg instead of 0.000001 kg. It is convenient however to have fractional or multiple units, and these are denoted by placing a prefix before the symbol of the unit.

It is normal to restrict the multiples and fractions of a unit to powers of 1000 (hence the centimetre, which is 10^{-2} m, is not strictly in accord with SI). The allowed prefixes are the following:

Fraction	Prefix	Symbol	Multiple	Prefix	Symbol
10^{-3}	milli	m	10^3	kilo	k
10^{-6}	micro	μ	10^6	mega	M
10^{-9}	nano	n	10^9	giga	G
10^{-12}	pico	p	10^{12}	tera	T
10^{-15}	femto	f			
10^{-18}	atto	a			

Compound prefixes should not be used, e.g. 10^{-9} metre is represented by 1 nm *not* 1 mμm.

When a prefix is attached to a unit, there is in effect a new unit. Thus, for example:

$$1\,\text{km}^2 = 1(\text{km})^2 = 10^6\,\text{m}^2$$
$$not \qquad 1\text{k(m)}^2 = 10^3\,\text{m}^2$$

Also note that until such time as a new name may be adopted for the kilogramme (the basic unit of mass) the gramme will often be used, both as an elementary unit (to avoid the absurdity of mkg) and in association with numerical prefixes, e.g., μg.

3.2 Relation of SI to CGS units

Because the move towards the standardization of units is a fairly recent one, many of the best books (books we may recommend you to consult) have been written in the rival system of units based on the *Centimetre*, *Gramme* and *Second*. You will, therefore, have to master the relationship between SI and CGS units. This affects most seriously the study of electricity and magnetism so here we compare the electrical units:

SI	CGS
1 A	10^{-1} e.m.u. ($= c/10$ e.s.u.)*
1 V	10^8 e.m.u. ($= 10^8/c$ e.s.u.)
1 Ω	10^9 e.m.u. ($= 10^9/c^2$ e.s.u.)
1 F	10^{-9} e.m.u. ($= 10^{-9}c^2$ e.s.u.)
1 H	10^9 e.m.u. ($= 10^9/c^2$ e.s.u.)
1 T	10^4 e.m.u. (gauss)
1 Wb	10^8 e.m.u. (maxwell)
1 A m^{-1}	$4\pi \times 10^{-3}$ e.m.u. (oersted)

($c \approx 3 \times 10^{10}$ in the above equations.)

* *There are* two *systems of CGS units: the* electromagnetic units (e.m.u.) *and the* electrostatic units (e.s.u.).

3.3 Other units

Finally we append an assorted list of common units that run contrary to SI, together with their SI equivalent:

Physical quantity	Unit	Equivalent
length	inch	0.025 4 m
	foot	0.304 8 m
	yard	0.914 4 m
	mile	1.609 34 km
	nautical mile	1.853 18 km
area	square inch	645.16 mm^2
	square foot	0.092 903 m^2
	square yard	0.836 127 m^2
	square mile	2.589 99 km^2
volume	cubic inch	$1.638\ 71 \times 10^{-5}$ m^3
	cubic foot	0.028 316 8 m^3
	U.K. gallon	0.004 546 092 m^3
mass	pound (avoirdupois)	0.453 592 37 kg
	slug	14.593 9 kg
density	pound/cubic inch	$2.767\ 99 \times 10^4$ kg m^{-3}
	pound/cubic foot	16.018 5 kg m^{-3}
force	dyne	10^{-5} N
	poundal	0.138 255 N
	pound-force	4.448 22 N
	kilogramme-force	9.806 65 N
pressure	atmosphere	101.325 kN m^{-2}
	torr	133.322 N m^{-2}
	pound-force/sq. in.	6 894.76 N m^{-2}
energy	erg	10^{-7} J
	calorie (I.T.)	4.186 8 J
	calorie (15°C)	4.185 5 J
	calorie (thermochemical)	4.184 J
	B.t.u.	1 055.06 J
	foot poundal	0.042 140 1 J
	foot pound-force	1.355 82 J
power	horse power	745.700 W
temperature	degree Rankine	(5/9) K
	degree Fahrenheit	°F = (9/5)°C + 32

Section 4

Dimensions

4.1 Derived SI units

In Section 3.1.3 we listed a number of *derived* SI units with special names.

For example, the unit of frequency, the hertz, was defined as

$$1 \text{ Hz} = 1 \text{ s}^{-1}$$

In the same list, you will find the unit of force, the newton, defined as

$$1 \text{ N} = 1 \text{ kg m s}^{-2}$$

If you are reading these notes *after* you have studied Unit 3 and Unit 4, you should certainly know why $1 \text{ N} = 1 \text{ kg m s}^{-2}$. If not, skip the next question for the time being.

Why is $1 \text{ N} = 1 \text{ kg m s}^{-2}$?

Force is *defined* as rate of change of *momentum*. Momentum is the product of mass and velocity; so if a mass of 1 kg increases its velocity by 1 m s^{-1} in 1 s we have unit increase of momentum in unit time and that is, by definition, what we call unit force.

Thus $1 \text{ N} = 1 \text{ kg} \times 1 \text{ m s}^{-1} \times 1 \text{ s}^{-1}$
$= 1 \text{ kg m s}^{-2}$

4.2 Derived SI units, unit symbols and dimensions

A simpler example of a derived unit, which was not listed in section 3.1.3 because it has no special name, is that of *area*, the square metre, written 1 m^2.

By way of an exercise, see if you can write in the SI symbols for the quantities named and defined in the following list:

Physical quantity	SI unit and definition	Symbol for SI unit	Dimensions
area	square metre	m^2	
velocity	metre per second	m s^{-1}	
momentum	kilogramme metre per second	kg m s^{-1}	
force	kilogramme metre per second per second (newton)	kg m s^{-2}	
volume	cubic metre		
density	kilogramme per cubic metre		
acceleration	metre per second squared (rate of increase of velocity)		
pressure	newton per square metre (force per unit area)		
dynamic viscosity	newton second per square metre		

Look at the quantities in the list above.

We may say that an *area*, which is obtained by multiplying a *length* by another *length* has the *dimensions of length squared*.

A *velocity*, which is obtained by dividing a *distance* by a *time*, has the dimensions of distance per time or distance $\times$ time^{-1}.

A shorthand way of saying the same thing is:

$$[\text{area}] = [\text{length}^2] = [L^2]$$

$$[\text{velocity}] = [\text{distance/time}] = [LT^{-1}]$$

And you can easily extend the idea, for instance:

$$[\text{momentum}] = [\text{mass} \times \text{velocity}] = [MLT^{-1}]$$

$$[\text{acceleration}] = [\text{velocity/time}] = [LT^{-2}]$$

and so on.

As a simple exercise, enter the appropriate *dimensions* in the fourth column of the table above.

4.3 The use of dimensions to check formulae

One obvious use of dimensions is as a check on whether you are using the right formula for something. For instance, suppose you want to calculate the distance travelled in a given time by a uniformly accelerated body (such as the ball in the Galileo experiment in Unit 1).

You realize that the distance must depend on the acceleration and on the time, but you are not sure whether it goes as t or t^2 or t^3. Suppose you try t^3. Then you say

$$[\text{distance}] = \text{some constant or other} \times [\text{acceleration}] \times [\text{time}^3]$$

$$[L] = [LT^{-2}] \times [T^3] \text{ (the constant is a pure number, it has no}$$
$$= [LT], \qquad\qquad \text{dimensions)}$$

which is obviously wrong; there should not be a T on the right hand side.

Evidently you should have had $[\text{time}^2]$ to start off with.

This is, of course, a trivial example. But it is often very useful, when you are working out a complicated formula, to check whether the dimensions on the two sides of the equality sign really are the same.

Suppose, for instance, you looked up a formula for the volume of liquid flowing per second through a tube of radius a and length l, there being a pressure difference P between its ends.

You have been given the following formula

$$V = \frac{P\pi a^3}{8\eta l} \quad \text{m}^3\,\text{s}^{-1}$$

with the constant η being described as the dynamic viscosity in N s m^{-2}.

Can you check whether the formula is right?

Check the dimensions:

$$[P] = [\text{Force/Area}] = [MLT^{-2}/L^2]$$
$$= [ML^{-1}T^{-2}]$$
$$[a] = [\text{Length}] = [L]$$
$$[\eta] = [\text{Force} \times \text{time/length squared}]$$
$$= [MLT^{-2} \times T/L^2] = [ML^{-1}T^{-1}]$$
$$[l] = [\text{Length}] = [L]$$

4.4 Dimensions and the relationship between physical quantities

Another use of dimensions is as a means of guessing how a set of physical quantities may be related, and thereby how one of them might change if you changed another.

Take, for example, the problem of a simple pendulum, consisting of a bob on the end of a string. We might guess that the period of oscillation (the time for one swing) of the pendulum could depend on the length of the string and the weight of the bob.

Now the weight of the bob is its mass times the acceleration due to gravity. So we are saying:

Period of pendulum = some constant × [length]^x × [mass]^y

× [acceleration]^z

where we have put in x, y and z for the unknown powers of the three physical quantities, length, mass and acceleration, on which we think the period might depend.

In dimensional terms, this could be written:

$$[T] = [L]^x \times [M]^y \times [LT^{-2}]^z$$

or

$$[L]^0[M]^0[T]^1 = [L]^{x+z} \times [M]^y \times [T]^{-2z}$$

This can only be true if

$-2z = 1$ (giving us $[T]^1$ on the right hand side, to agree with the left)

So

$z = -\frac{1}{2}$

$\left.\begin{array}{l} y = 0 \\ x + z = 0 \end{array}\right\}$ (since there is no $[L]$ or $[M]$ on the left hand side, i.e. they are there to the power zero)

So

$x = \frac{1}{2}$

Thus we have found:

Period of pendulum = some constant × (length)$^{\frac{1}{2}}$ × (acceleration)$^{-\frac{1}{2}}$

Or, more briefly

$$T_0 = C\sqrt{\frac{l}{g}}$$

where C is a dimensionless constant, to be determined by calculation or experiment.

Note that we have found that the period does not depend on the weight of the bob.

So

$$\left[\frac{P\pi a^3}{8\eta l}\right] = \left[\frac{ML^{-1}T^{-2} \times L^3}{ML^{-1}T^{-1} \times L}\right]$$

$$= \left[\frac{ML^2T^{-2}}{MT^{-1}}\right] = [L^2T^{-1}]$$

But it should have the dimensions L^3T^{-1} (i.e., volume per second). We have to multiply by something with the dimensions of length. This could be done by leaving out the l from the denominator. But that would make the flow rate independent of the length of the tube, which is contrary to experience and to common sense.

Another way of getting the dimensions right would be to change a^3 to a^4. Yet another way would be to change a^3 to a^5 and l to l^2.

Note that the dimensional method of checking a formula can never be completely without ambiguity. And note also that it cannot give you any idea of what value a dimensionless constant, like 8π, should have.

As it happens, our guess before last was the right one in this particular case, and the formula should be

$$V = \frac{P\pi a^4}{8\eta l} \text{ m}^3 \text{ s}^{-1}$$

43

Self-assessment Questions

The following tests are to enable you to gain additional practice in the use of the various formulae quoted in *The Handling of Experimental Data*. Answers are to be found on pp. 47–50.

Question 1

The diameter of a wire is measured repeatedly in different places along its length. The measurements in millimetres are:

$$1.26, 1.26, 1.29, 1.31, 1.28, 1.27, 1.26, 1.25,$$

$$1.28, 1.32, 1.21, 1.27, 1.22, 1.29, 1.28$$

Calculate the standard deviation of these readings.

Question 2

Calculate s_m, the error on the mean, for the data listed in question 1.

Question 3

Calculate the standard deviation of the data given in question 1 using the simplified expression, equation (6).

Question 4

Two objects have weights (100 ± 0.4) g and (50 ± 0.3) g. (a) What is the error on the sum of their weights? (b) What is the error on the difference between their weights? The individual errors may be assumed independent.

Question 5

Two objects are weighed and each is found to be 200 g to within an accuracy of ± 0.5 per cent. What percentage accuracy can be assigned to the sum of the weights?

Question 6

Three objects have weights (100 ± 0.4) g, (50 ± 0.3) g, and (200 ± 0.5) g. What is the error on their sum?

Question 7

A motorist drives at a constant speed such that the reading on the speedometer is 40 mph. The speedometer is assumed to be accurate to ± 2 mph. At the end of the morning he would like to know how far he has travelled but unfortunately he forgot to look at the mileage indicator when he set out. He reckoned that he had been driving for 4 hours, give or take a quarter of an hour. Estimate how far he travelled and assign an error to your result.

Question 8

The volume of a rectangular block is calculated from the following measurements of its dimensions: (10.00 ± 0.10) cm, (5.00 ± 0.05) cm, and (4.00 ± 0.04) cm. What is the error on the value of the volume of the block assuming (a) the errors are independent, (b) the errors are correlated such that they all push the estimate in the same sense?

Question 9

The pressure of a gas can be estimated from the force exerted on a given area. If the force is (20.0 ± 0.5) newtons, and the area is rectangular with sides (5.0 ± 0.2) mm and (10.0 ± 0.5) mm, what is the fractional error on the value of the pressure?

Question 10

What is the error on the value of the area of a circle whose radius is determined to be (14.6 ± 0.5) cm?

Question 11

The area of a circle is given by: area $= \pi r^2$ where r is the radius. This can be written: area $= (\pi r) \times r$. Can we regard this equation as being similar in form to that of equation 12 where $Z =$ area, $A = \pi r$, and $B = r$, and hence use equation 13 instead of 16 for determining the error on the area?

Question 12

The time period, T, of a simple pendulum is given by the expression $T = 2\pi \sqrt{\dfrac{l}{g}}$, where l is the length of the pendulum, and g is the acceleration due to gravity. A pendulum of length (0.600 ± 0.002) m is used to determine the value of g. The value of T was found to be (1.55 ± 0.01) s. Find the percentage error on the value of g.

Question 13

Given the following table of values, calculate a value for the slope of the straight line graph through them by the points-in-pairs method. Also determine the values of x and y for one point through which you would draw the graph.

x	2.0	4.0	6.0	8.0	10.0	12.0
y	7.1	8.2	8.8	10.0	11.0	11.8

Question 14

In a certain experiment the total pressure exerted on a gas was the pressure, P, recorded on a gauge plus the unknown atmospheric pressure, B. Corresponding values of P and the volume, V, were recorded. On the assumption that the total pressure was inversely proportional to V, P was plotted against $(1/V)$. The points were found to be on a straight line. The points-in-pairs approach gave the slope to be 1.10×10^6 kN m, the mean values of P and $(1/V)$ were respectively $\bar{P} = 215$ kN m^{-2}, and $\overline{(1/V)} = 2.85 \times 10^{-4}$ m^{-3}. By estimating the value of an intercept, find B.

Questions on Section 2, "A Guide to Practical Work"

	50-point Check List item number
Try the following questions before turning to the appropriate item in the 50-point Check List in section 2.15 (p. 34) for the answer.	
How is the mean value of several repeated readings calculated?	2
How is the standard deviation of the sample calculated?	3
What percentage of the readings are expected to lie within one standard deviation of the mean?	3
How can an estimate be made of the standard error on the mean of several readings?	4
Name at least three items that might enter an experiment as a systematic error.	5
Given several contributing errors how is the resultant error calculated?	6
Give an expression for the error	7
on a sum	7
on a difference	8
on a product	9
on a quotient	10
on an exponential.	11
How big does a contributing error have to be before it is necessary to include it in the calculation of the resultant error?	12
If $y = Kx^3$, where K is a constant, what variables would you plot?	14
How is the best straight line by the method of points-in-pairs obtained?	15
How are the intercepts on the axes obtained?	16
Which variable, the dependent or independent one, should be plotted along a horizontal axis, and what is this axis called?	17
Which of the following symbols is/are acceptable when plotting points on a graph: **+ · ⊙ ?**	20
What should appear as labels on the axes of a graph?	21
Should the readings be *always* evenly spread out over the range of values?	23
What three points should you bear in mind when recording the measurements?	26, 27, 28
How are wrongly written numbers to be corrected in an account?	30
How should you go about checking a calculation for an arithmetical slip?	34
How can you avoid giving the impression that certain noughts are significant figures when in fact they are only there to indicate the position of the decimal point?	37
When three numbers variously known to 3, 5, and 9 significant figures are multiplied, how many significant figures has the product?	38
What considerations should be borne in mind when composing the title of a final report?	40
Should an abstract *always* be included in a final report? What should an abstract contain and how long should it be?	42
What should immediately follow an abstract?	43
What sort of items should be included in the section on conclusions?	48

Self-assessment Answers and Comments

Answer 1

Answer: $s \approx 0.03$ mm (actually 0.028 mm)

Hint: first calculate the mean using equation 1. This gives $\bar{x} = 1.27$ mm. Then calculate the residuals from equation 2, square them, add the squares, divide by 15 to get the mean, and take the square root as in equation 3.

Answer 2

$s = 0.028, n = 15$

$\therefore$ From equation 4

$$s_m = \frac{0.028}{(15 - 1)^{\frac{1}{2}}} = \frac{0.028}{3.74}$$

Answer:

$$s_m = 0.0075 \text{ mm}$$

As the original readings were only quoted to two places after the decimal point, it would be in order to round this value up to 0.01 mm and quote the value of $\bar{x}$ and its error as:

$$\bar{x} = (1.27 \pm 0.01) \text{ mm}$$

Answer 3

Subtracting the mean value, 1.27 mm, from each reading, gives the following list for the moduli of the residuals: 0.01, 0.01, 0.02, 0.04, 0.01, 0, 0.01, 0.02, 0.01, 0.05, 0.06, 0, 0.05, 0.02 and 0.01 mm. The mean value of these is obtained by taking the sum and dividing by 15:

$$r = 0.32/15 = 0.0213 \text{ mm}.$$

$\therefore$ From equation 6 $\quad s \approx \dfrac{5 \times 0.0213}{4} = 0.0266 \text{ mm}.$

This compares quite well with the previous value of 0.028 mm obtained from equation 3.

Answer 4

(a) From equation 10:

$$\text{error} = \{(0.4)^2 + (0.3)^2\}^{\frac{1}{2}}$$
$$= (0.16 + 0.09)^{\frac{1}{2}}$$
$$= (0.25)^{\frac{1}{2}}$$
$$= 0.5 \text{ g}$$

(b) Equation 10 still applies

$$\therefore \text{ error} = 0.5 \text{ g}$$

Answer 5

$$\text{The error on each} = \frac{0.5}{100} \times 200 = 1 \text{ g}$$

$$\therefore \text{ Error on sum} = (1^2 + 1^2)^{\frac{1}{2}} = 1.4 \text{ g}$$

$$\therefore \text{ Percentage error} = \frac{1.4}{200 + 200} \times 100 = \underline{0.35\%}$$

Answer 6

When there are more than two quantities the same rule applies: sum the squares of the individual errors and take the square root.

$$\therefore \text{ Error on the sum} = \{(0.4)^2 + (0.3)^2 + (0.5)^2\}^{\frac{1}{2}}$$

$$= (0.16 + 0.09 + 0.25)^{\frac{1}{2}}$$

$$= (0.50)^{\frac{1}{2}}$$

$$= \underline{0.7 \text{ g}}$$

Answer 7

$x = vt$, where x = distance travelled, v = speed, and t = time.

$$\therefore x = 40 \times 4 = \underline{160 \text{ miles}}$$

$$\Delta v = \pm 2 \text{ mph} \quad \Delta t = \pm 0.25 \text{ hour}$$

$\therefore$ From equation (13)

$$\frac{\Delta x}{x} = \left\{ \left(\frac{2}{40}\right)^2 + \left(\frac{0.25}{4}\right)^2 \right\}^{\frac{1}{2}} = \{(0.05)^2 + (0.0625)^2\}^{\frac{1}{2}}$$

$$= (0.0025 + 0.0039)^{\frac{1}{2}} = (0.0064)^{\frac{1}{2}} = 0.08$$

$$\therefore \Delta x = 0.08 \times x = 0.08 \times 160$$

$$\therefore \underline{\Delta x = 12.8 \text{ miles}}$$

Answer 8

$V = abc$, where V = volume, and a, b, and c, are the dimensions

$$\therefore V = 10 \times 5 \times 4 = 200 \text{ cm}^3$$

(a) $$\frac{\Delta V}{V} = \left\{ \left(\frac{\Delta a}{a}\right)^2 + \left(\frac{\Delta b}{b}\right)^2 + \left(\frac{\Delta c}{c}\right)^2 \right\}^{\frac{1}{2}}$$

$$= \left\{ \left(\frac{0.1}{10}\right)^2 + \left(\frac{0.05}{5}\right)^2 + \left(\frac{0.04}{4}\right)^2 \right\}^{\frac{1}{2}}$$

$$= \{0.01^2 + 0.01^2 + 0.01^2\}^{\frac{1}{2}}$$

$$= \{3 \times 0.01^2\}^{\frac{1}{2}} = \sqrt{3} \times 0.01 = 0.017$$

$$\therefore \Delta V = 0.017 \times 200 = \underline{3.4 \text{ cm}^3}$$

(b) In this case we just add the fractional errors without first squaring them

$$\therefore \frac{\Delta V}{V} = 0.01 + 0.01 + 0.01 = 0.03$$

$$\therefore \Delta V = 0.03 \times 200 = \underline{6 \text{ cm}^3}$$

Answer 9

$P = \dfrac{F}{a \times b}$ where P = pressure, F = force, a and b are the lengths of the sides of the area. As the same rule applies for finding the errors on ratios as on products, the same rule will apply to a mixture of ratio and product.

$$\therefore \frac{\Delta P}{P} = \left\{ \left(\frac{\Delta F}{F}\right)^2 + \left(\frac{\Delta a}{a}\right)^2 + \left(\frac{\Delta b}{b}\right)^2 \right\}^{\frac{1}{2}}$$

$$= \left\{ \left(\frac{0.5}{20}\right)^2 + \left(\frac{0.2}{5}\right)^2 + \left(\frac{0.5}{10}\right)^2 \right\}^{\frac{1}{2}}$$

$$= \left\{ \left(\frac{0.5}{20}\right)^2 + \left(\frac{0.8}{20}\right)^2 + \left(\frac{1.0}{20}\right)^2 \right\}^{\frac{1}{2}}$$

$$= \frac{(0.25 + 0.64 + 1.00)^{\frac{1}{2}}}{20}$$

$$= \frac{(1.89)^{\frac{1}{2}}}{20} = \frac{1.37}{20}$$

Answer: $\dfrac{\Delta P}{P} = 0.0685$

Answer 10

$Z = \pi r^2$ where Z = area, r = radius

$$Z = \pi \times 14.6^2$$

From equation 16

$$\frac{\Delta Z}{Z} = \frac{2\Delta r}{r}$$

$$\therefore \frac{\Delta Z}{\pi \times 14.6^2} = \frac{2 \times 0.5}{14.6}$$

$$\therefore \Delta Z = 1 \times \pi \times 14.6$$

Answer: $\Delta Z = 46 \text{ cm}^2$

Answer 11

If we were to use equation 13 we would get

$$\frac{\Delta Z}{Z} = \left\{ \left(\frac{\Delta(\pi r)}{(\pi r)}\right)^2 + \left(\frac{\Delta r}{r}\right)^2 \right\}^{\frac{1}{2}}$$

$$= \left\{ \left(\frac{\Delta r}{r}\right)^2 + \left(\frac{\Delta r}{r}\right)^2 \right\}^{\frac{1}{2}}$$

$$\frac{\Delta Z}{Z} = \sqrt{2}\frac{\Delta r}{r}$$

But from equation 16 we had

$$\frac{\Delta Z}{Z} = 2\frac{\Delta r}{r}$$

Therefore something must be wrong. The fallacy is that in equation 13 it is assumed that the errors on A and B are *independent*. When $A = (\pi r)$ and $B = r$ this is clearly not so; a certain fractional increase in B leads to the same fractional increase in A; the effect on Z produced by an error on B is always enhanced by the correlated error on A and there can be no partial cancellation. Thus equation 13 gives an underestimate for the error and must not be used in circumstances such as these.

Answer 12

Rearranging the equation into a form that allows g to be calculated from l and T.

$$T = 2\pi\sqrt{l/g} \qquad T^2 = 4\pi^2 l/g$$

$$g = \frac{4\pi^2 l}{T^2}$$

The fractional error on g is given by equation 13

$$\frac{\Delta g}{g} = \left[\left(\frac{\Delta l}{l}\right)^2 + \left(\frac{\Delta(T^2)}{T^2}\right)^2\right]^{\frac{1}{2}}$$

But from equation 16 (putting $Z = T^2$)

$$\frac{\Delta(T^2)}{T^2} = 2\frac{\Delta T}{T}$$

$$\therefore \frac{\Delta g}{g} = \left[\left(\frac{\Delta l}{l}\right)^2 + \left(\frac{2\Delta T}{T}\right)^2\right]^{\frac{1}{2}}$$

$$= \left[\left(\frac{2}{600}\right)^2 + \left(\frac{2 \times 0.01}{1.55}\right)^2\right]^{\frac{1}{2}}$$

$$= [(0.003)^2 + (0.013)^2]^{\frac{1}{2}} = (0.09 + 1.69)^{\frac{1}{2}} \times 10^{-2}$$

$$= 1.35 \times 10^{-2}$$

$\therefore$ Percentage error on $g = 1.35 \times 10^{-2} \times 100\%$

$\therefore$ Answer $= \underline{1.35\%}$

Answer 13

The values of x are equally spaced so all we have to do is to divide the six y values into two groups of three and subtract the first y value from the fourth, the second from the fifth, and the third from the sixth:

$$(y_4 - y_1) = 10.0 - 7.1 = 2.9$$
$$(y_5 - y_2) = 11.0 - 8.2 = 2.8$$
$$(y_6 - y_3) = 11.8 - 8.8 = 3.0$$

The mean difference, $\bar{D}_y = \dfrac{2.9 + 2.8 + 3.0}{3} = 2.9$

The equal spacing of these pairs of points is 6.0, therefore from equation 22, the slope is given by

$$m = \frac{\bar{D}_y}{D_x} = \frac{2.9}{6.0} = 0.48$$

The line is to be drawn through the point $(\bar{x}, \bar{y})$ where

$$\bar{x} = \frac{2.0 + 4.0 + 6.0 + 8.0 + 10.0 + 12.0}{6} = 7.0$$

and

$$\bar{y} = \frac{7.1 + 8.2 + 8.8 + 10.0 + 11.0 + 11.8}{6} = 9.5$$

Answer: Slope $= \underline{0.48}$

The point has $x = 7.0$, $y = 9.5$

Answer 14

The total pressure $(P + B)$ is inversely proportional to V

$\therefore\ P + B = m(1/V)$

i.e.

$$P = m(1/V) - B$$

where m is the constant of proportionality. This equation is in the form $y = mx + c$

where

$y = P, x = (1/V)$ and $c = -B$

In the text it is shown that the intercept on the y axis is c. $\therefore$ In this particular experiment the intercept on the P axis must be equal to $-B$. Using $c = \bar{y} - m\bar{x}$, we have

$$-B = 215 - (1.10 \times 10^6) \times (2.85 \times 10^{-4})$$
$$-B = 215 - 314 = -99$$
$$\therefore\ \underline{B = 99\ \text{kN m}^{-2}}$$

S.100—SCIENCE FOUNDATION COURSE UNITS